Statistical Downscaling for Hydrological and Environmental Applications

Statistical Downscaling for Hydrological and Environmental Applications

Taesam Lee and Vijay P. Singh

CRC Press is an imprint of the
Taylor & Francis Group, an **informa** business

CRC Press
Taylor & Francis Group
6000 Broken Sound Parkway NW, Suite 300
Boca Raton, FL 33487-2742

Printed on acid-free paper

International Standard Book Number-13: 978-1-138-62596-9 (Hardback)

Visit the Taylor & Francis Web site at
http://www.taylorandfrancis.com

and the CRC Press Web site at
http://www.crcpress.com

Printed and bound in Great Britain by
TJ International Ltd, Padstow, Cornwall

TL: Wife Misun and children Sooin and Yeojun

VPS: Wife Anita, son Vinay, daughter Arti, daughter-in-law Sonali, son-in-law Vamsi, and grandsons Ronin, Kayden, and Davin

Contents

Preface

Although climate change is understood as a gradual variation at large spatial and long temporal scales, real impacts of climate change are experienced on short-space time scales, such as extreme rainfall and floods, and some impacts are also experienced on large-space time scales, such as droughts, hurricanes, heat waves, cold waves, extreme snowfall, and winds. Global climate change is understood and modeled using global climate models (GCMs), but the outputs of these models in terms of hydrological variables are only available on coarse or large spatial and time scales, but finer spatial and temporal resolutions are needed to reliably assess hydroenvironmental impacts of climate change. To obtain required resolutions of hydrological variables, statistical downscaling has been popularly employed.

In the field of hydrologic science and engineering, a number of statistical downscaling techniques have been reported during the past several decades. Much of the literature on statistical downscaling is available in journals, dissertations, and technical reports, which are not always easily accessible. As a result, their applications seem limited to only those who are well versed in the area of downscaling. A book that solely deals with statistical downscaling and covering available approaches and techniques seems to be in high demand.

The major objective of this book therefore is to present statistical downscaling techniques in a practical manner so that students and practitioners can easily adopt the techniques for hydrological applications and designs in response to climate change. The techniques are illustrated with examples.

The subject matter of the book is divided into nine chapters. Chapter 1 introduces the concept of statistical downscaling and its necessity and also discusses the coverage and focus of the book. Chapter 2 provides the basic statistical background that is employed for statistical downscaling approaches. Chapter 3 explains the type of datasets that are required for statistical downscaling and how and where the datasets can be acquired. Since GCM outputs often present significant biases with observational data, Chapter 4 describes how these biases can be corrected.

In statistical downscaling, according to the purpose of climate change assessments, final outputs and datasets employed are quite different from each other. In case a weather station-based hydrological output is needed, a pointwise downscaling method is necessary. In Chapter 5, transfer function downscaling methods, mostly based on regression type approaches, are described as a pointwise method.

Regression-type models rely on the dependence between climate variables and surface hydrological variables, also called transfer function model. However, surface hydrological variables, such as precipitation and streamflow, do not often exhibit significant relationships with climate variables, even though the long-term evolution of these variables is affected by climate variables. Therefore, stochastic simulation models that only relate climate variables to the parameters of stochastic simulation models have been developed for hydrological variables. Different weather generator models employing stochastic simulation schemes are therefore presented in Chapter 6.

Another downscaling approach is weather typing by classifying atmospheric circulation patterns to a limited number of weather types. After selecting a weather type for a certain day, stochastic weather simulation can be adopted by conditioning the defined weather type to simulate the local variable of interest for future climate. The procedure is described in Chapter 7.

High temporal resolution of climate variables (e.g., precipitation) is critical for the assessment of extreme hydrological events, such as floods. Chapter 8 describes a temporal downscaling method. Meanwhile, downscaling approaches to obtain spatially coarser datasets are explained in Chapter 9.

The book will be useful to graduate students, college faculty, and researchers in hydrology, hydroclimatology, agricultural and environmental sciences, and watershed management who are interested in climate change and its hydrologic impacts. It may also be useful to policymakers in government at local, state, and national levels.

The first author acknowledges his university (Gyeongsang National University) supporting his sabbatical fund for this book and Texas A&M University for providing an excellent environment. Also, he expresses gratitude to the Asia-Pacific Economic Cooperation climate center in Busan motivating to write this book for research officials from developing countries.

Taesam Lee
Vijay P. Singh
College Station, Texas

List of Abbreviations

ANN	Artificial Neural Network
AR	Autoregression
AR5	IPCC's Fifth Assessment Report
ASE	Approximation Square Error
BC	Bias Correction
BCCA	Bias Correction and Constructed Analogues
BCSA	Bias Correction and Stochastic Analogues
BCSD	Bias Correction and Spatial Downscaling
CDF	Cumulative Distribution Function
CMIP5	Coupled Model Intercomparison Project, Phase 5
CORDEX	Coordinate Regional Climate Downscaling Experiment
CV	Cross Validation
ECDF	Empirical Cumulative Distribution Function
GA	Genetic Algorithm
GCM	Global Climate Model
GEV	Generalized Extreme Value
KNN	K-Nearest Neighbor
KNNR	K-Nearest Neighbor Resampling
LASSO	Least Absolute Shrinkage and Selection Operator
LWT	Lamb Weather Type
MAR	Multivariate Autoregressive Model
MCS	Monte Carlo Simulation
MLE	Maximum Likelihood Estimation
MOM	Method of Moments
NCAR	National Center for Atmospheric Research
NCEP	National Centers for Environmental Prediction
NTD	Nonparametric Temporal Downscaling
NWGEN	Nonparametric Weather Generator
PDF	Probability Density Function
QDM	Quantile Delta Mapping
QM	Quantile Mapping
RCM	Regional Climate Model
RCP	Representative Concentration Pathways
RMSE	Root Mean Square Error
SWR	Stepwise Regression
WGEN	Weather Generator

Authors

Taesam Lee, Ph.D., is a professor in the Department of Civil Engineering at Gyeongsang National University. He got his Ph.D. degree from Colorado State University with stochastic simulation of streamflow and worked at Institut National de la Recherche Scientifique Centre Eau Terre Environnement (INRS-ETE) in Quebec City as a research fellow with Prof. Taha B.M.J. Ouarda. He specializes in surface-water hydrology, meteorology, and climatic changes in hydrological extremes publishing more than 40 technical papers.

Vijay P. Singh, Ph.D., D.Sc., D. Eng. (Hon.), Ph.D. (Hon.), P.E., P.H., Hon. D. WRE is a distinguished professor, a regents professor, and Caroline & William N. Lehrer distinguished chair in water engineering in the Department of Biological and Agricultural Engineering and Zachry Department of Civil Engineering at Texas A&M University. He specializes in surface-water hydrology, groundwater hydrology, hydraulics, irrigation engineering, and environmental and water resources engineering. He has published extensively and has received more than 90 national and international awards.

1 Introduction

1.1 WHY STATISTICAL DOWNSCALING?

Recent studies have shown that atmospheric concentration of greenhouse gases, such as carbon dioxide, has significantly increased due to human activities. Greenhouse gases have led to regional and global changes in climate and climate-related variables, such as temperature and precipitation (Watson et al., 1996). Climate models have the ability to predict the consequences of the increasing concentration of greenhouse gases.

However, resolutions of climate models are quite coarse for hydrologic and environmental modeling that impact assessments through mathematical models (e.g., drought/flood modeling) for which employing global climate model (GCM) output variables is not suitable. To bridge this gap, regional climate models (RCMs) are developed, based on dynamic formulations using initial and time-dependent lateral boundary conditions of GCMs, with no feedback from the RCM simulation to the driving of GCM (IPCC, 2001). The basic RCM techniques are essentially from numerical weather prediction originally developed by Dickinson et al. (1989) that provide a spatial resolution up to 10–20 km or less. However, these RCMs are still not fine enough to provide the detailed spatial–temporal scale outputs needed for small watershed hydrological and field agricultural climate impact studies, where local and site-specific scenarios are required (Chen et al., 2011). This might hinder extensive applications of GCM projections for the impact assessment of climate change.

Statistical downscaling methods have been developed to specify the GCM or RCM scenarios into further fine spatial and temporal scales employing statistical relations between climate output variables of the scenarios and the variables of local interest. The statistical downscaling methods are computationally cheaper and much easier to implement than RCMs. Statistical downscaling is to establish statistical links between large-scale atmospheric patterns and observed local-scale weather to resolve the scale discrepancy between climate change scenarios and the resolution for impact assessment at a watershed scale or a station level (Maraun et al., 2010).

After establishing a statistical relation from large-scale atmospheric patterns in climate models and weather variables of local interest, future projections of local interest variables can be made with the established relation and future atmospheric patterns of climate models. The relationship between the current large-scale patterns and local weather must remain valid, called stationary. However, it is uncertain that the established statistical links will be upheld in the future climate system under different forcing conditions of possible future climates (Caffrey and Farmer, 2014). This is one of the weak points of statistical downscaling.

1.2 CLIMATE MODELS

Climate models can be categorized into three groups according to their complexity (McGuffie and Henderson-Sellers, 2005): (1) energy balance models, (2) intermediate complexity models (e.g., statistical dynamical models, earth system models of intermediate complexity, and integrated assessment models), and (3) global circulation models or GCMs.

Energy balance models are zero or one-dimensional models that estimate surface temperature as a function of energy balance of the earth with solar radiation. The observed solar radiation, which excludes the reflected solar radiation from solar radiation input, is equal to the emitted terrestrial radiation as

$$\pi R^2(1-\alpha_p)S_0 = 4\pi R^2 \sigma T_e^4 \tag{1.1}$$

where S_0 is the solar energy (1,370 W/m^2), and R and T^e are the earth radius (6,371 km) and the effective temperature of the earth, respectively. α_p and σ are the planetary albedo (0.3) and the Stefan–Boltzmann constant (5.67×10^{-8} W/m^2/K^4), respectively. Note that the emitted terrestrial radiation is the emitted energy from the earth $\left(\sigma T_e^4\right)$ by Stefan–Boltzmann law times the surface area of the spherical earth ($4\pi R^2$). From this balance, the effective temperature that equalizes the absorbed solar energy and the terrestrial radiation becomes

$$T_e \pi R^2 = \left[\frac{1}{4\sigma}(1-\alpha_p)S_0\right]^{1/4} \tag{1.2}$$

The effective temperature from this calculation becomes 255 K (= −18°C). However, the existence of atmosphere plays a vital role as greenhouse gases absorb thermal radiation, i.e., $T = T_e + \Delta T$, where T is the surface temperature and ΔT is the greenhouse increment. This variation is conceptually modeled in energy balance models.

Radiative–convective models explicitly calculate the fluxes of solar and thermal radiation for a number of layers in the atmosphere focused on the global average surface temperature. Dimensionally constrained climate models commonly represent the vertical and horizontal dimension combining the latitudinal dimension of the energy balance model plus the vertical radiative–convective model. GCMs calculate the full three-dimensional characteristics of the atmosphere and ocean.

GCMs have been employed to project the response of the global climate system to increase greenhouse gas concentrations. The fundamental equations of GCMs are derived from the basic laws of physics, such as the conservation laws for momentum, mass, and energy, governing the motions of the atmosphere, oceans, and sea ice on earth (Washington and Parkinson, 2005). GCMs present the global climate system with a three-dimensional grid with a horizontal resolution between 250 and 600 km and 10 and 20 vertical layers.

1.3 STATISTICAL DOWNSCALING

Statistical downscaling methods have normally been categorized into three parts, such as regression downscaling or transfer function model (Wilby et al., 2002), weather generator (Wilby et al., 1998), and weather typing (Bardossy and Plate,

1992; Storch et al., 1993). The transfer function techniques employ statistical linear or nonlinear relationships between observed local variables and GCM (or RCM) output variables. Instead of GCM outputs, a reanalysis dataset in which observations and a GCM are combined to generate a synthesized estimate of the state of the system is required to fit the relationships. The scenarios from GCMs do not account for the real past weather conditions. The transfer function techniques are normally applied when a specific site needs to assess the impact of climate change. The main drawback of this technique is the lack of statistical relationships for some variables (e.g., precipitation). When a weak relation or no relation is found, this technique cannot be applied. This is one of the main reasons that weather generators have been developed.

Weather generators employ stochastic simulation techniques with the perturbation of parameters that summarize the statistical characteristics of observations. The parameters are perturbed according to the changes derived by climate models for future conditions. A number of climate scenarios, including natural variability, for climate-related variables of interest (e.g., temperature and precipitation) at a specific site or multiple sites can be produced within a short computation time. The drawback of weather generators is that an infinite number of series can be simulated, which make it difficult to select a representative scenario among them; and specific climate conditions represented by climate models are indirectly delivered through the parameters of weather generators.

Weather-type approaches classify atmospheric circulation patterns into a number of types and relate the patterns to local meteorological variables (mainly precipitation). A future scenario of local variables is simulated by taking a classified pattern similar to the global circulation pattern from a GCM scenario. Therefore, the local variables are closely linked to global circulation (Bardossy and Plate, 1992). However, the major drawbacks are (1) no strong relation can frequently be found between atmospheric circulation patterns and local meteorological variables (especially precipitation) and (2) future circulation patterns may significantly deviate from observations not contained in the past observations.

1.4 SELECTION OF MODEL SCHEME

Defining which statistical downscaling method is used belongs to which output data is needed. If the future change of a hydrometeorological variable (e.g., temperature and precipitation) according to climate scenarios is needed for assessing climate change impacts at a specific site, the bias-correction in Chapter 4 is sometimes good enough to apply. When the future evolution of a local variable (especially temperature) affected by climate variables is needed, the regression-based downscaling methods in Chapter 5 with the procedure illustrated in Figure 1.1 can be applied. A regression model is established with climate variables of the reanalysis data (explained in Chapter 3) and a variable of local interest. The future evolution of variables of local interest is predicted by the climate variables of GCM scenarios. When hydrometeorological variables are too random and significant, relations with other variables (e.g., temperature) cannot be obtained, and weather generator (stochastic downscaling) discussed in Chapter 6 with the procedure

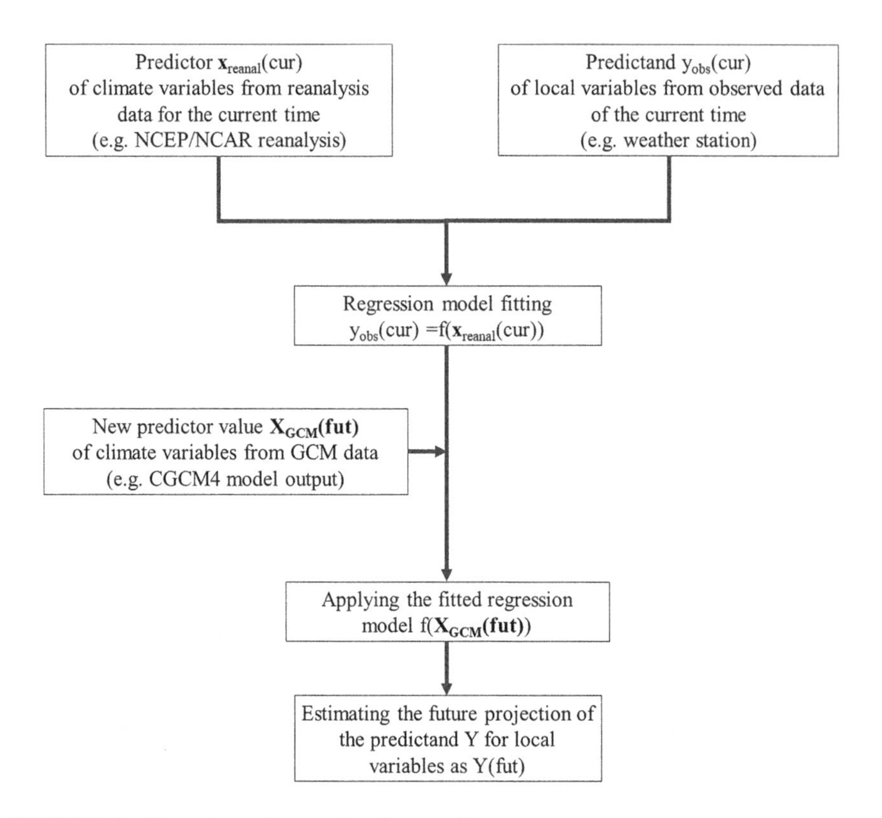

FIGURE 1.1 Procedure of regression downscaling.

illustrated in Figure 1.2 should be applied. After fitting a stochastic model to the observed data of variables of interest, future climate scenarios of variables of interest are simulated by updating the model parameter set defined with the GCM scenarios. A physical process is adapted in a statistical downscaling method in a way that a large-scale atmospheric circulation is mapped to a small number of categories to condition local-scale precipitation (Figure 1.3) described in Chapter 7. After conditioning weather types, a stochastic simulation model is applied in Chapter 6.

When a fine temporal resolution of GCM or RCM outputs is needed (e.g., hourly precipitation for small watersheds or urban areas), temporal downscaling discussed in Chapter 8 with the procedure illustrated in Figure 1.4 can be used. The bias-corrected GCM or RCM output can be applied to a predefined temporal downscaling model with the observed coarse or fine time scale data. If the spatial resolution of GCM outputs is too coarse to apply in a hydrological simulation, the spatial downscaling discussed in Chapter 9 with the procedure illustrated in Figure 1.5 can be used. Observed gridded data with coarse and spatial resolutions are employed to establish a spatial downscaling model, and the bias-corrected coarse GCM grid data is downscaled to the fine GCM grid data.

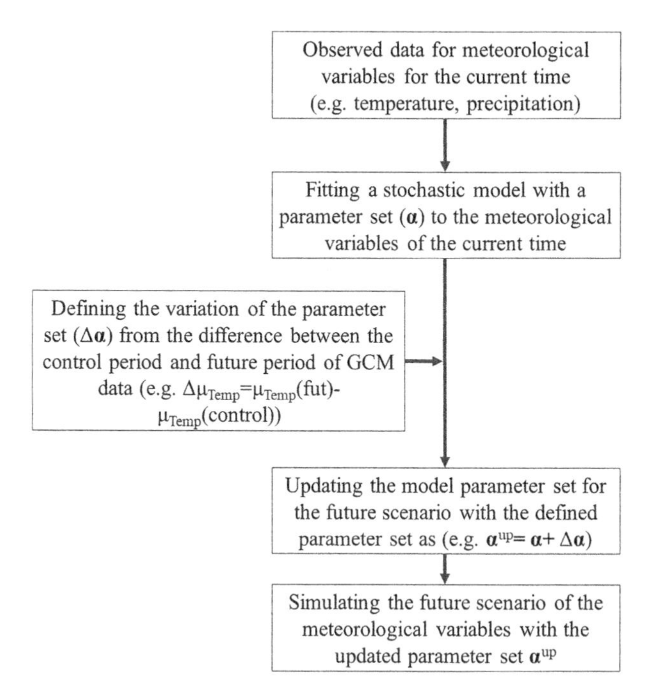

FIGURE 1.2 Procedure of weather generator downscaling (or stochastic downscaling).

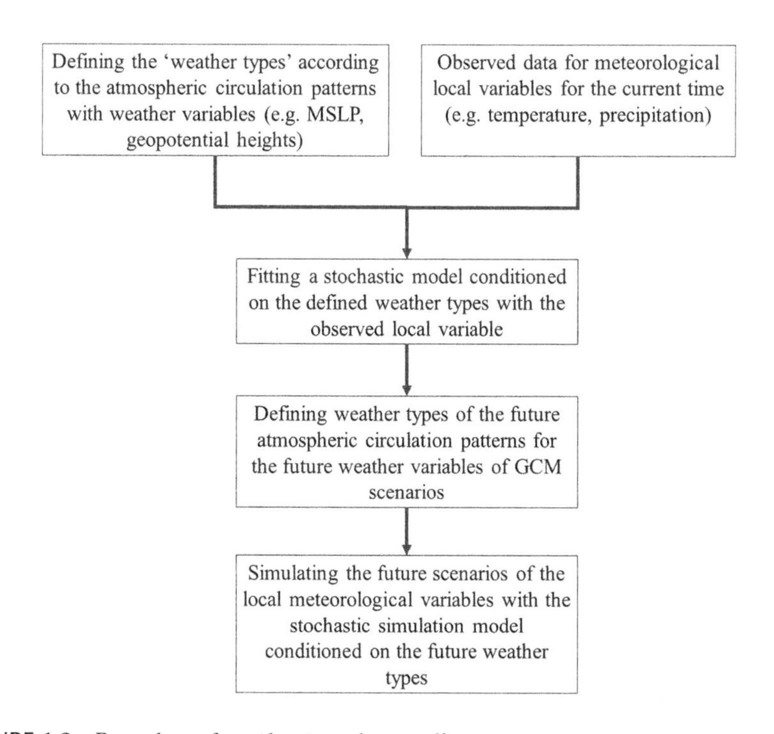

FIGURE 1.3 Procedure of weather-type downscaling.

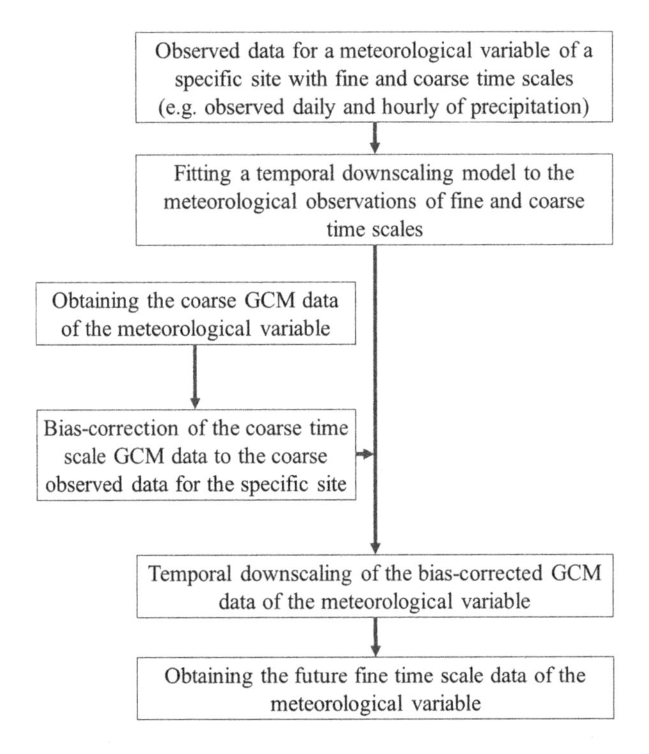

FIGURE 1.4 Procedure of temporal downscaling.

1.5 STRUCTURE OF CHAPTERS

The chapters of this book are based on the objectives of statistical downscaling. Statistical background and data and their formats description are given in Chapters 2 and 3. Chapter 4 discusses bias correction, since bias correction is mainly applied to the outputs of climate models and the bias-corrected outputs of climate models are then employed for downscaling. The regression downscaling is explained in Chapter 5 for both linear and nonlinear models. In linear regression downscaling, predictor selection methods are also added, since the selection of key variables among a number of output variables of climate models is critical and the selection methods are applicable for linear regression models. Weather generator (stochastic downscaling) is discussed in Chapter 6. Weather-type downscaling is presented in Chapter 7. The three approaches discussed in Chapters 5–7 are mostly used to obtain site-specific future climate scenarios of local variables.

In contrast, Chapters 8 and 9 deal with temporal and spatial downscaling approaches, respectively. These chapters mainly focus on precipitation, which is fundamental for hydrological and agricultural applications. Temporal downscaling of daily to hourly precipitation is especially presented, since it is critical for small watersheds and daily precipitation outputs are available from recent GCM and RCM models. Spatial downscaling describes methods that produce fine grid data from coarse GCM models.

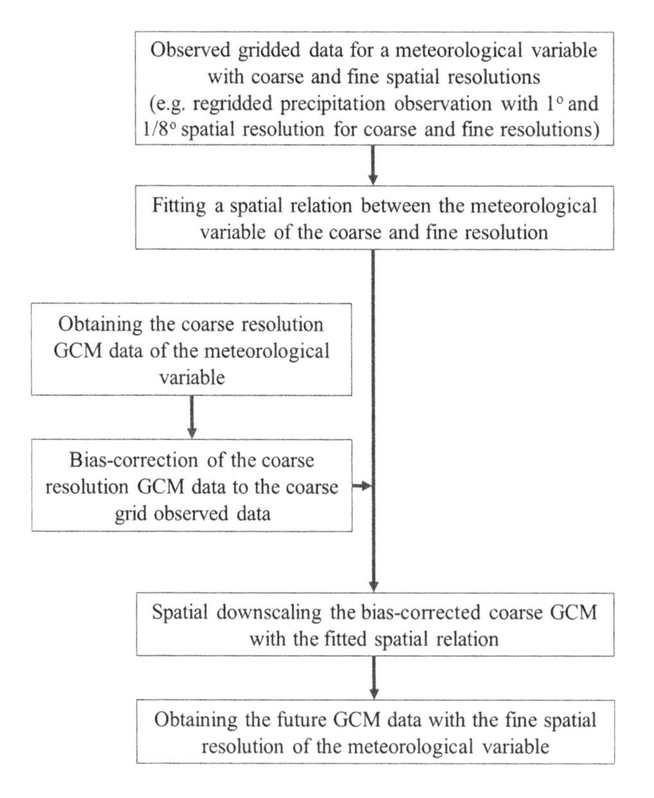

FIGURE 1.5 Procedure of spatial downscaling.

1.6 SUMMARY AND CONCLUSION

Statistical downscaling methods have been popularly developed in the last few decades for impact assessment of climate change in various fields, such as agriculture, hydrology, and environment, since the produced outputs from climate models have a significant scale discrepancy. Even though individual investigations have been overwhelming, an appropriate book, that systematically describes different methodologies of statistical downscaling is lacking and is needed. Also, a number of software for statistical downscaling are available, but their fundamental backgrounds are described in a limited way or not easily understandable. Therefore, the main objective of this book is to present detailed explanations of statistical downscaling approaches according to their purposes of application. Instead of providing detailed theoretical backgrounds of the statistical approaches, the focus of the book is on how to implement statistical downscaling in practice and explain and illustrate them with examples.

This book would be beneficial for undergraduate and graduate students who are interested in the assessment of hydrologic and environmental impacts of climate change. For beginners in governmental and educational sectors who need to assess the impacts of climate change, this book might serve as a guide to become familiar with computational procedures.

2 Statistical Background

2.1 PROBABILITY AND STATISTICS

2.1.1 Probabilistic Theory

2.1.1.1 Probability Density Function and Cumulative Distribution Function

A random variable is a variable whose possible values depend on a set of random outcomes and is a function that maps probability to a physical outcome, denoted as a capital italic letter (e.g., X). A random variable X can be either continuous or discrete. For a random variable, an event for a specific value x can be defined as $X = x$, $X > x$, or $X < x$.

For a random variable X, its probability distribution can be described by the cumulative distribution function (CDF), denoted as

$$F_X(x) = P(X \leq x) \tag{2.1}$$

For a discrete random variable, its probability distribution may also be described as probability mass function (PMF), denoted as

$$p_X(x_i) = P(X = x_i) \text{ and } F_X(x) = \sum_{all\ x_i \leq x} P(X = x) \tag{2.2}$$

where x_i is one of the possible outcomes of the discrete random variable.

For a continuous random variable, there is no probability for a specific value x, i.e., $P(X = x) = 0$. Therefore, the probability distribution is described in terms of probability density function (PDF) denoted as

$$f_X(x) = \frac{dF_X(x)}{dx} \tag{2.3}$$

2.1.1.2 Descriptors of Random Variables

Statistical characteristics of a random variable, such as mean, variance (or standard deviation), skewness, and kurtosis, can be described from a number of measurements. For example, the mean value describes the central value of the range of a random variable. The mean of a continuous random variable, μ, can be denoted as

$$E(X) = \mu = \int_{-\infty}^{\infty} x f_X(x) dx \tag{2.4}$$

where $E[X]$ is the expectation of the random variable X.

The value of the random variable with the largest probability or the highest probability density is called the *mode* (denoted as $\tilde{x}$) and can be found at the place of $df_X(x)/dx = 0$, because the largest place indicates the maximum point of a PDF with zero inclination. Furthermore, the value at which the CDF is 50% is called the median (denoted as x_m), i.e., $F_X(x_m) = 0.5$.

To measure the dispersion of a random variable, variance (σ^2) and standard deviation (σ) are popularly employed. The variance (*Var*) is the second central moment and can be denoted as

$$Var(X) = \sigma^2 = E[(X-\mu)^2] = \int_{-\infty}^{\infty} (x-\mu)^2 f_X(x)dx \tag{2.5}$$

The standard deviation is the square root of the variance as $\sigma = \sqrt{Var(X)}$. To present the degree of dispersion more meaningfully, the coefficient of variation, δ_X, defined as the ratio of the standard deviation to the mean is expressed as

$$\delta_X = \frac{\sigma_X}{\mu_X} \tag{2.6}$$

The symmetry of a PDF is a useful descriptor. A measure of this asymmetry, called skewness, is

$$E\left[(X-\mu)^3\right] = \int_{-\infty}^{\infty} (x-\mu)^3 f_X(x)dx \tag{2.7}$$

The coefficient of skewness γ is defined as

$$\gamma = \frac{E\left[(X-\mu)^3\right]}{\sigma^3} \tag{2.8}$$

The fourth central moment of a random variable, known as the *kurtosis*, is a measure of the peakedness of the underlying PDF of X expressed as

$$E\left[(X-\mu)^4\right] = \int_{-\infty}^{\infty} (x-\mu)^4 f_X(x)dx \tag{2.9}$$

Note that the kurtosis for a standard normal distribution is three. The coefficient of kurtosis g is defined as

$$g = \frac{E\left[(X-\mu)^4\right]}{\sigma^4} \tag{2.10}$$

2.1.2 Discrete Probability Distributions

Probability distributions can be separated into discrete and continuous for the characteristics of the corresponding random variable (i.e., discrete or continuous r.v.).

There are a number of probability distributions for each category, and only a few important distributions are introduced here. Most common discrete probability distributions are Bernoulli and Binomial, among others.

2.1.2.1 Bernoulli Distribution

In a single trial of an experiment with two possible outcomes (e.g., rain or no rain for a day), the probability distribution of two outcomes is called a Bernoulli distribution, and its trial is called a Bernoulli trial. The outcomes are also considered as occurrence or nonoccurrence, i.e., 1 or 0, respectively. It is assumed that the occurrence probability (i.e., $p = p(X = 1)$, $0 \le p \le 1$) is constant, and it is obvious that the nonoccurrence probability (i.e., $p(X = 0) = 1 - p$), since the outcomes are mutually exclusive.

The PMF of the Bernoulli distribution is

$$\begin{aligned} p_X(x) &= p^x(1-p)^{1-x} && \text{for } x = 0,1 \\ &= 0 && \text{otherwise} \end{aligned} \tag{2.11}$$

Its mean and variance can be described as follows:

$$E(X) = 0 \times (1-p) + 1 \times p = p \tag{2.12}$$

$$\begin{aligned} Var(X) = E[X - E(X)]^2 = E[X-p]^2 &= (0-p)^2(1-p) + (1-p)^2 p \\ &= p(1-p)\{p + (1-p)\} = p(1-p) \end{aligned} \tag{2.13}$$

2.1.2.2 Binomial Distribution

Binomial distribution is applied for an experiment of n successive Bernoulli trials that are independent of each other and whose occurrence probability ($p = p(X = 1)$, $0 \le p \le 1$) is constant. The outcome (X) is the number of successes (or occurrences) among n trials. The binomial PMF is

$$\begin{aligned} p_X(x) &= \binom{n}{x} p^x(1-p)^{n-x} && \text{for } x = 0,1,\ldots,n \\ &= 0 && \text{otherwise} \end{aligned} \tag{2.14}$$

where $\binom{n}{x} = \frac{n!}{x!(n-x)!}$ is the total number of possible combinations when choosing x objects from n objects. Its mean and variance are mathematically calculated as

$$E(X) = np \tag{2.15}$$

$$Var(X) = E[X - E(X)]^2 = np(1-p) \tag{2.16}$$

2.1.3 Continuous Probability Distributions

There are a number of probability distributions for continuous random variables. The continuous probability distributions can be categorized as symmetric and asymmetric distributions especially for hydrological variables. When a probability distribution exhibits a certain value (δ) such that $f(\delta + x) = f(\delta - x)$, then the distribution is symmetric. In other words, the probabilities of two values ($\delta + x$ and $\delta - x$) with the same distance from the certain value (δ) are always the same. However, probability distributions of hydrological variables, such as precipitation and streamflow, are often asymmetric, because these variables range from 0 to infinity and have no negative values. It is important to adopt appropriate distributions for hydrological variables.

The most common symmetric distribution is normal (or Gaussian). The standard normal distribution that standardizes (subtracting mean and dividing by standard deviation) is commonly employed. Exponential, lognormal, and gamma distributions are frequently employed for representing hydrological variables, such as precipitation amount and streamflow. Furthermore, some distributions, such as Gumbel and generalized extreme value (GEV), are commonly employed for representing extreme hydrological events, such as droughts and floods. A few of the commonly employed continuous distributions are presented in what follows.

2.1.3.1 Normal Distribution and Lognormal Distributions

The normal distribution, also known as Gaussian distribution, is the most widely employed probability distribution. Its PDF for a random variable X is

$$f_X(x) = \frac{1}{\sigma\sqrt{2\pi}} \exp\left(-\frac{(x-\mu)^2}{2\sigma^2}\right) \quad -\infty < x < \infty \tag{2.17}$$

where μ and σ are the parameters of this distribution represented by mean and variance in Eqs. (2.4) and (2.5). Here, μ plays the role of locating the center of the normal distribution, while σ plays the role of its dispersion (i.e., higher σ means higher variability of the corresponding random variable).

With the parameters of $\mu = 0$ and $\sigma = 1$, the distribution is known as the standard normal distribution. A variable with the normal distribution (i.e., $X \sim N(\mu, \sigma)$) can be transformed as

$$Z = \frac{X - \mu}{\sigma}$$

Then, the variate Z becomes a standard normal, whose pdf is

$$f_Z(z) = \frac{1}{\sqrt{2\pi}} \exp\left(-\frac{z^2}{2}\right) \tag{2.18}$$

Sometimes it is also referred to as error distribution or function. The CDF of the standard normal distribution is

$$F_Z(z) = \int_{-\infty}^{z} \frac{1}{\sqrt{2\pi}} \exp\left(-\frac{t^2}{2}\right) dt = \Phi(z) \tag{2.19}$$

The normal distribution is shown in Figure 2.1. While the thick solid line presents the PDF curve of the standard normal distribution, the center-shifted PDF of the normal distribution with $\mu = 1$ and $\sigma = 1$ (i.e., $N(1, 1)$) from the standard normal distribution (i.e., $N(0,1)$) is shown with the dotted line with x markers as well as the normal distribution with variance increased as much as two times of the standard normal (i.e., $N(0,2)$) with dash-dotted line with + markers. The CDF of standard normal distribution is illustrated in Figure 2.2 for the fixed value z.

The lognormal distribution is the logarithmic version of the normal distribution, whose PDF is

$$f_X(x) = \frac{1}{\zeta x \sqrt{2\pi}} \exp\left(-\frac{(x-\varsigma)^2}{2\zeta^2}\right) \tag{2.20}$$

where $\zeta = E(\ln X)$ and $\varsigma = \sqrt{Var(\ln X)}$ are the parameters of the distribution.

2.1.3.2 Exponential and Gamma Distributions

The PDF and CDF of the exponential distribution are

$$f_E(x) = \frac{1}{\alpha} \exp\left(-\frac{x}{\alpha}\right) \tag{2.21}$$

$$F_E(x) = 1 - \exp\left(-\frac{x}{\alpha}\right) \tag{2.22}$$

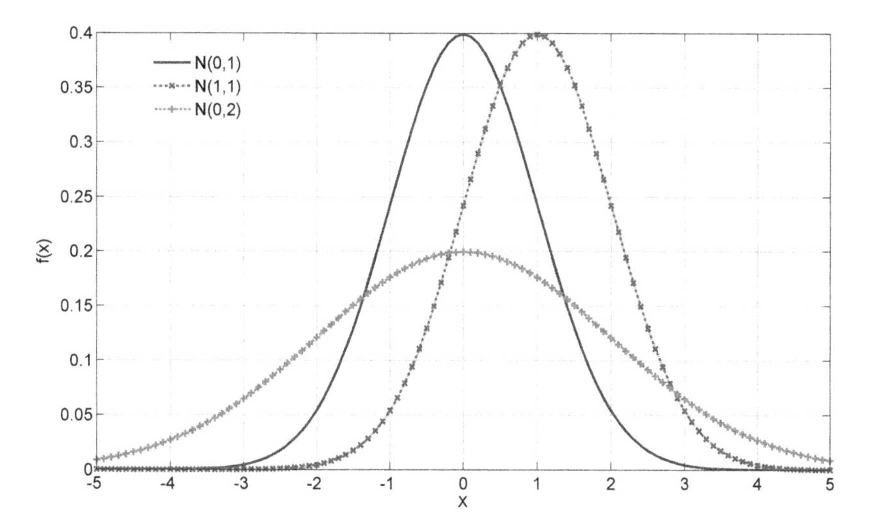

FIGURE 2.1 PDFs of a normal distribution for $N(0,1)$ with thick solid line as well as the center-shifted normal distribution (i.e., $N(1,1)$, dotted line with x markers) and the variance increased normal distribution (i.e., $N(0,2)$, dot-dashed line with + markers).

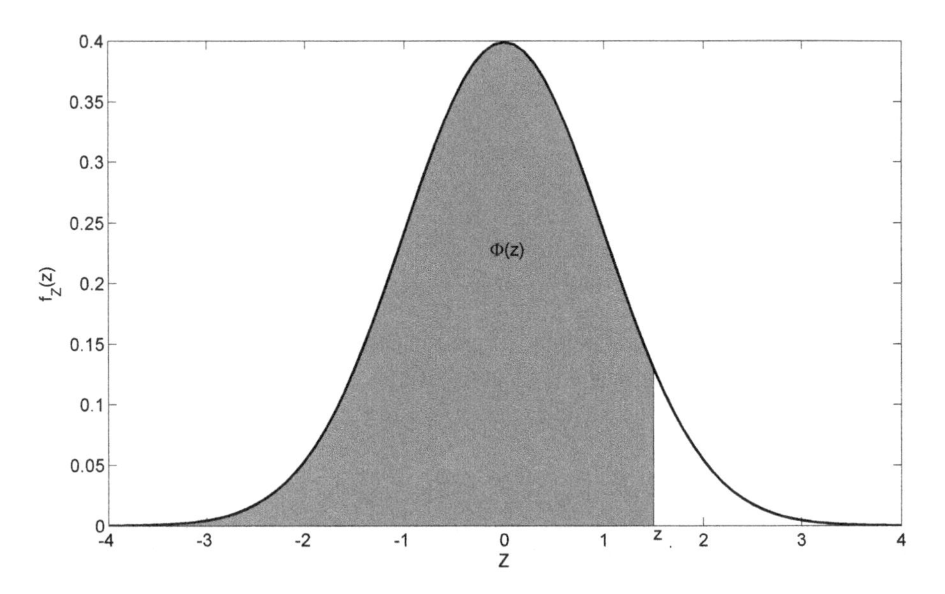

FIGURE 2.2 PDF of standard normal distribution (black thick line) and CDF for a fixed z value (shaded area).

where α is the parameter of exponential distribution. Its quantile can also be estimated as

$$x = -\alpha \ln\{1 - F_E(x)\} \tag{2.23}$$

Its expectation or mean can be derived as

$$E(X) = \mu = \int_0^\infty x f_E(x) dx = \alpha \tag{2.24}$$

In general, the gamma distribution has been widely used to fit RCM data and observed data, especially for the precipitation variable (Piani et al., 2010). Its PDF and CDF are

$$f_G(x) = \frac{\beta^{-\alpha}}{\Gamma(\alpha)} x^{\alpha-1} \exp\left(-\frac{x}{\beta}\right) \tag{2.25}$$

$$F_G(x) = \int_0^x \frac{\beta^{-\alpha}}{\Gamma(\alpha)} x^{\alpha-1} \exp\left(-\frac{x}{\beta}\right) dx;\ x, a, b > 0 \tag{2.26}$$

where $\Gamma(\cdot)$ is the gamma function, and α and β are the shape and scale parameters, respectively. Its expectation and variance have the following relationship with these parameters:

$$E(X) = \mu = \int_0^\infty x f_G(x) dx = \alpha\beta \tag{2.27}$$

$$Var(X) = E[X - E(X)]^2 = \sigma^2 = \alpha\beta^2 \tag{2.28}$$

Here, parameters α and β can be estimated from an observed dataset and various methods, such as method of moments (MOM), maximum likelihood, probability-weighted moments, L-moments, amongst others.

The distribution parameters can be easily estimated with these two equations as

$$\beta = \frac{\sigma^2}{\mu} \tag{2.29}$$

$$\alpha = \frac{\mu}{\beta} \tag{2.30}$$

Three-parameter gamma distribution, known as Pearson type III distribution or shifted gamma distribution, has been popularly employed in hydrologic field and can be defined as

$$f_G(x) = \frac{\beta^{-\alpha}}{\Gamma(\alpha)}(x - \gamma)^{\alpha-1} \exp\left(-\frac{x - \gamma}{\beta}\right) \tag{2.31}$$

2.1.3.3 Generalized Extreme Value and Gumbel Distribution

The PDF and CDF of the GEV distribution for a random variable X can be expressed as

$$f(x) = \frac{1}{\alpha}\left[1 - \beta\left(\frac{x - x_0}{\alpha}\right)\right]^{(1/\beta)-1} F(x) \tag{2.32}$$

$$F(x) = \exp\left\{-\left[1 - \beta\left(\frac{x - x_0}{\alpha}\right)\right]^{1/\beta}\right\} \tag{2.33}$$

where α, β, and x_0 are the scale, shape, and location parameters, respectively. β plays an important role such that if $\beta = 0$, the distribution tends to resemble an extreme type-1 or Gumbel distribution; if $\beta < 0$, the resulting distribution is extreme value type-2 or log-Gumbel distribution; and if $\beta > 0$, it is extreme type-3 distribution or Weibull distribution. The quantile function for GEV and Gumbel distributions corresponding to the nonexceedance probability q is given in Eqs. (A12) and (A13), respectively, as

$$x_T = x_0 + \frac{\alpha}{\beta}\left\{1 - \left[-\ln\left(1 - \frac{1}{T}\right)\right]^{\beta}\right\} \text{ for GEV} \tag{2.34}$$

$$x_T = x_0 - \alpha \ln\left[-\ln\left(1 - \frac{1}{T}\right)\right] \text{ for Gumbel} \tag{2.35}$$

where x_T is the quantile of return period T for GEV and Gumbel distributions.

2.1.4 Parameter Estimation for Probability Distributions

The parameters of probability distributions may be estimated with observations obtained from population and represented by "samples of the population." The exact values of the population parameters are generally unknown, because the population cannot be observed in general and only finite samples can be. Thus, the parameters must be estimated with the observed finite samples. There are different estimation methods, such as MOM and maximum likelihood estimation (MLE).

2.1.4.1 Method of Moments

The mean and variance given by Eqs. (2.4) and (2.5) are the first moment and the central second moment, respectively, and are the most common among other moments. These moments are related to the parameters of probability distributions. For example, the mean of the Bernoulli distribution is directly related with the parameter p as shown in Eq. (2.12). Statistics estimated with a set of sample data may be used to replace the moments and related with the parameters. The sample mean and sample variance are defined with sample data of $x_1,\ldots, x_n$ as

$$\bar{x} = \frac{1}{n}\sum_{i=1}^{n} x_i \tag{2.36}$$

$$s^2 = \frac{1}{n-1}\sum_{i=1}^{n}(x_i - \bar{x})^2 \tag{2.37}$$

2.1.4.2 Maximum Likelihood Estimation

The rationale underlying MLE is to find the parameter set that maximizes the likelihood (or the highest possibility) for the observed sample data to have come from the population. The likelihood with the assumption of random sampling (i.e., independent observations) is defined as the multiplication of the PDFs, given the parameter set:

$$L(x_1,\ldots,x_n \mid \boldsymbol{\theta}) = f(x_1|\boldsymbol{\theta})f(x_1|\boldsymbol{\theta})\ldots f(x_n|\boldsymbol{\theta}) = \prod_{i=1}^{n} f(x_i|\boldsymbol{\theta}) \tag{2.38}$$

where L is the likelihood function and $\boldsymbol{\theta}$ is the parameter set.

The parameter set maximizing the likelihood in Eq. (2.38) can be found by taking partial derivatives for each parameter and equal to zero as

$$\frac{\partial L(x_1,\ldots,x_n|\boldsymbol{\theta})}{\partial \theta_k} = 0 \quad k = 1,\ldots,M \tag{2.39}$$

where M is the number of parameters of the probability distribution. In practice, log-likelihood (LL) is commonly used, since the multiplication of the PDF is too small to calculate, and the logarithm monotonically transforms the likelihood function as a summation:

$$LL(x_1,\ldots,x_n|\boldsymbol{\theta}) = \sum_{i=1}^{n} \log\left(f(x_i|\boldsymbol{\theta})\right) \tag{2.40}$$

Example 2.1

Fit the exponential distribution given in Eq. (2.21) with MOM or MLE to the January precipitation dataset at Cedars, Quebec, presented in the second column of Table 2.1.

Solution:

The parameter estimation with the MOM is rather straightforward, since the mean is the same as the exponential parameter ($\alpha = \mu$) as in Eq. (2.24). The sample mean can be estimated as Eq. (2.36):

$$\bar{x} = \frac{39.4 + 64.3 + \cdots + 68.3}{26} = 60.042$$

Therefore, the MOM estimate is $\hat{\alpha} = 60.042$. Note that the hat over the parameter α indicates the estimate from the sample data.

TABLE 2.1

January Precipitation (mm) at Cedars, Quebec, as well as Its Order, the Increasing-Ordered *X*, and Plotting Positions in Eq. (2.49)

	Jan Precip. (mm)	Order (i)	Increasing Ordered X	Plotting Positions $i/(n+1)$
1961	39.4	1	19.0	0.037
1962	64.3	2	19.8	0.074
1963	58.7	3	30.0	0.111
1964	110.7	4	30.5	0.148
1965	41.1	5	31.9	0.185
1967	84.9	6	36.3	0.222
1968	56.0	7	39.4	0.259
1969	107.8	8	40.0	0.296
1970	31.9	9	41.1	0.333
1971	86.4	10	44.4	0.370
1972	36.3	11	47.3	0.407
1973	30.0	12	53.6	0.444
1974	30.5	13	54.2	0.481
1975	47.3	14	56.0	0.519
1976	84.8	15	58.7	0.556
1977	86.1	16	64.3	0.593
1978	96.5	17	68.3	0.630
1979	54.2	18	72.7	0.667
1980	19.8	19	84.8	0.704
1981	19.0	20	84.9	0.741
1984	44.4	21	86.1	0.778
1985	40.0	22	86.4	0.815
1986	96.4	23	96.4	0.852
1987	72.7	24	96.5	0.889
1989	53.6	25	107.8	0.926
1990	68.3	26	110.7	0.963

For the MLE of the exponential distribution, the LL can be denoted as

$$LL(x_1, x_2, \ldots, x_n \mid \alpha) = -\sum_{i=1}^{n} \log\left\{\frac{1}{\alpha}\exp\left(-\frac{x_i}{\alpha}\right)\right\} = \sum_{i=1}^{n}\left(\log\alpha + \frac{x_i}{\alpha}\right)$$

To find the maximum of LL, the derivative of the LL is taken and set to be zero as

$$\frac{\partial LL}{\partial \alpha} = \sum_{i=1}^{n}\left(\frac{1}{\alpha} - \frac{x_i}{\alpha^2}\right) = 0$$

$$\frac{n}{\alpha} = \sum_{i=1}^{n}\frac{x_i}{\alpha^2} \text{ and } \alpha = \sum_{i=1}^{n}\frac{x_i}{n} = \bar{x} \tag{2.41}$$

This derivation indicates that the MLE estimate of the exponential distribution is the same as the MOM estimate.

Example 2.2

Estimate the parameters of the gamma distribution in Eq. (2.25) with MLE.

Solution:

First, the LL function is constructed as

$$LL(x_1, x_2, \ldots, x_n \mid \alpha, \beta) = \sum_{i=1}^{n} \log\left\{\frac{\beta^{-\alpha}}{\Gamma(\alpha)} x^{\alpha-1} \exp\left(-\frac{x_i}{\beta}\right)\right\}$$

$$= n(-\alpha\log(\beta) - \log(\Gamma(\alpha))) + (\alpha - 1)\sum_{i=1}^{n}\log(x_i) - \frac{1}{\beta}\sum_{i=1}^{n} x_i$$

Taking the partial derivatives of LL with respect to parameters, we obtain

$$\frac{\partial LL(\alpha, \beta \mid \mathbf{x})}{\partial \alpha} = -n\log(\beta) - n\frac{d}{d\alpha}\log(\Gamma(\alpha)) + \sum_{i=1}^{n}\log(x_i) = 0$$

$$\frac{\partial LL(\alpha, \beta \mid \mathbf{x})}{\partial \beta} = -\frac{1}{\beta}n\alpha + \frac{1}{\beta^2}\sum_{i=1}^{n} x_i = 0$$

Here, it is obvious from the second derivative that

$$\beta = \frac{\sum_{i=1}^{n} x_i}{n\alpha} = \frac{\bar{x}}{\alpha} \tag{2.42}$$

Substituting Eq. (2.42) into the first derivative given earlier, we get

$$-n\log(\bar{x}) - n\log(\alpha) + n\frac{d}{d\alpha}\log(\Gamma(\alpha)) + \sum_{i=1}^{n}\log(x_i) = 0 \tag{2.43}$$

Equation (2.43) must be solved numerically to estimate the parameter α.

2.1.5 Histogram and Empirical Distribution

Histogram is a graphical representation of a probability distribution with an observed dataset using bars of different heights. The following steps can be taken to draw a histogram, assuming that the dataset with n number of data, is arranged in increasing order as

1. Determine the range of the dataset as $x_{\min}$ and $x_{\max}$
2. Divide the range of the dataset into the number of classes (or bins), n_c, and set the base length Δx accordingly. A typical formula to estimate n_c (Sturges, 1926; Kottegoda and Rosso, 2008) is

$$n_c = 1 + 3.3\log_{10} n \tag{2.44}$$

where n is the number of data. The base length Δx can be estimated as

$$\Delta x = \frac{x_{\max} - x_{\min}}{n_c - 1} \tag{2.45}$$

3. Determine the number of data for the ith bin ($n_{(i)}, i = 1,\ldots,n_c$), whose range is defined as $R(i) = [x_{Low} + (i-1)\Delta x, x_{Low} + (i)\Delta x]$.
4. Draw a bar plot with the assigned number of data ($n_{(i)}$) for the y-axis and the corresponding data range ($R(i)$) for the x-axis.

The absolute frequency for the ith bin is then estimated as

$$\hat{p}(i) = \frac{n_{(i)}}{n} \quad \text{for } i = 1,\ldots,n_c \tag{2.46}$$

and its empirical CDF $F(i)$ as

$$\hat{F}(i) = \sum_{j=1}^{i} \hat{p}(j) = \sum_{j=1}^{i} \frac{n_{(j)}}{n} \quad \text{for } i = 1,\ldots,n_c \tag{2.47}$$

While the histogram is usually drawn by a bar plot, the top-center point of each bin can be connected, and it becomes a frequency polygon. Furthermore, the frequency polygon may be transformed into an *empirical density function* as

$$\hat{f}(i) = \frac{\hat{p}(i)}{\Delta x} = \frac{n_{(i)}}{n\Delta x} \tag{2.48}$$

Note that $\hat{p}(i) \overset{emp}{\approx} \Pr\left(x_{Low} + (i-1)\Delta x \le X \le x_{Low} + (i)\Delta x\right)$ and $\hat{f}(i) \overset{emp}{\approx} d\hat{p}(i)/dx$, where $\overset{emp}{\approx}$ denotes the empirical approximation.

For a set of increasing ordered n observations denoted as $x_{(1)}, x_{(2)},\ldots, x_{(n)},$ a probability paper can be constructed with their corresponding cumulative probabilities whose value (called plotting position) is determined as

$$\hat{F}\left(x_{(i)}\right)=\frac{i}{(N+1)} \tag{2.49}$$

where $x_{(i)}$ is the ith increasing-ordered value and N is the number of data. Though other plotting positions have been suggested, the plotting position in (2.49) is theoretically satisfactory as well proved by Gumbel (1954) and has been most widely employed. Also, Cunnane (1978) suggested a single compromise formula for all distributions as

$$\hat{F}\left(x_{(i)}\right)=\frac{(i-0.4)}{(N+0.2)} \tag{2.50}$$

which is also a commonly used formula.

Example 2.3

Estimate the empirical density function in Eq. (2.48) and draw the histogram for the January precipitation dataset at Cedars, Quebec in Table 2.1 with their corresponding years. Note that some data values are missing.

Solution:

To draw a histogram, the number of class (n_c) and the base length (Δx) must be determined after setting the minimum and maximum values of the dataset. Though the minimum and maximum of the dataset is 19.0 and 110.7 mm, it is better to give a little margin for both the minimum and maximum values of the histogram instead of using the observed minimum and maximum. Here, 10 and 120 mm are set for the minimum and maximum values of the histogram. The number of bins is now calculated as

$$n_c=1+3.3\log_{10}26=5.669$$

Since a natural number must be employed for the number of bins, set $n_c = 6$. The base length is

$$\Delta x=\frac{120-10}{6-1}=22$$

The following steps are taken to draw a histogram:

a. Set the data range for each bin (here, six bins are used) as in the first column of Table 2.2.
b. Count the number of data that are inside the range of each bin. For example, there are five values within 10–32 mm from the January precipitation in the second column of Table 2.1. All the counted numbers are given in the third column of Table 2.2.

TABLE 2.2
Datasets for Histogram and January Precipitation (mm) at Cedars, Quebec, as well as Its Order

Data Range	Meddian	No. of Data	Relative Frequency	Empirical Density Function	No. of Accumulated Data	ECDF
10 (min.)						
32	21	5[a]	0.192[b]	0.0087[c]	5	0.192
54	43	7	0.269	0.0122	12	0.461[d]
76	65	6	0.231	0.0105	18	0.692
98	87	6	0.231	0.0105	24	0.923
120 (max.)						
	109	2	0.077	0.0035	26	1.000
Sum		26	1.000			

[a] This is the number of data in the range of 10–32 mm, whose median is 21 mm for the data values in the second column of Table 2.1.

[b] Relative frequency is estimated by dividing the number of data for each bin (here, 5) with the total number of data (here, 26), i.e., 5/26 = 0.192.

[c] Empirical density function is estimated by dividing the relative frequency in the fourth column with the base length Δx (here, 22), i.e., 0.192/22 = 0.0087.

[d] ECDF is estimated by accumulating the relative frequency and its corresponding x value must be the maximum of each bin, i.e., 0.192 + 0.269 = 0.461 and its corresponding value is 54 mm, since this value indicates that $\hat{F}(i) = \Pr(X \le 54)$.

c. Estimate the relative frequency $\hat{p}(i)$ and the empirical density function $\hat{f}(i)$ as $n_{(i)}/n$ and $n_{(i)}/n\Delta x$, respectively, as given in the fourth and fifth columns of Table 2.2.

The histograms of the number of data (top panel), the relative frequency (middle panel), and the empirical density function (bottom panel) are presented in Figure 2.3. Note that the median value of each data range must be located at the center of each bin.

The empirical cumulative distribution function (ECDF) can be directly estimated with the accumulated number of data values for the bins as shown in the sixth column of Table 2.2 (see the solid line with circle markers). In addition, the plotting position can be calculated with $i/(n+1)$ (see the fifth column of Table 2.1) corresponding to the increasing-ordered dataset as shown in the fourth column of Table 2.1. These two cumulative probabilities are presented in Figure 2.4. Note that the ECDF has fewer values than the plotting position probabilities, and its last probability of the empirical cumulative distribution (ECD) is 1.0, which is not realistic [i.e., suppose that $\Pr(X \le 120) = 1.0$ and $\Pr(X > 120) = 0.0$]. Therefore, the cumulative probability of plotting position is more frequently employed in practice.

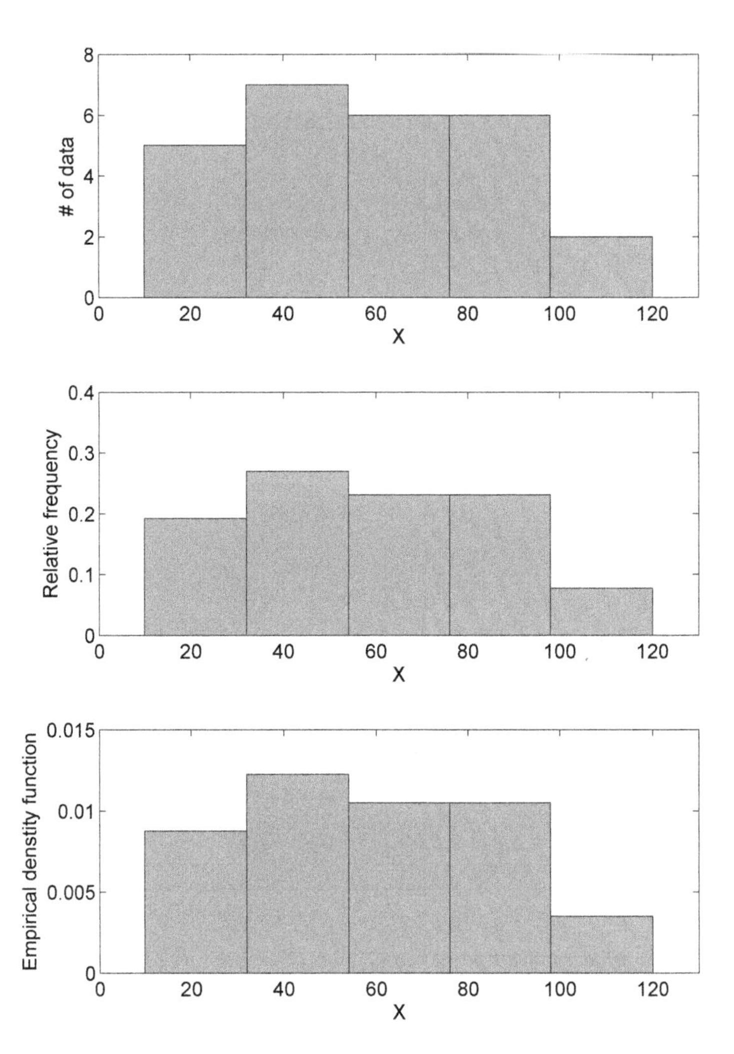

FIGURE 2.3 Histogram of the January precipitation data at Cedars, Quebec, for (a) the number of data values $n_{(i)}$ (top panel); (b) relative frequency $n_{(i)}/n$ (middle panel); and (c) empirical density function $n_{(i)}/n\Delta x$ (bottom panel).

2.2 MULTIVARIATE RANDOM VARIABLES

2.2.1 Multivariate Normal Distribution and Its Conditional Distribution

The p-dimensional normal density of the random vector $\boldsymbol{X} = [X_1, X_2, \ldots, X_p]^T$ is expressed as

$$f(\mathbf{x}) = \frac{1}{(2\pi)^{p/2}|\Sigma|^{1/2}} \exp\left[\frac{-(\mathbf{x}-\mu)^T \Sigma^{-1}(\mathbf{x}-\mu)}{2}\right] \tag{2.51}$$

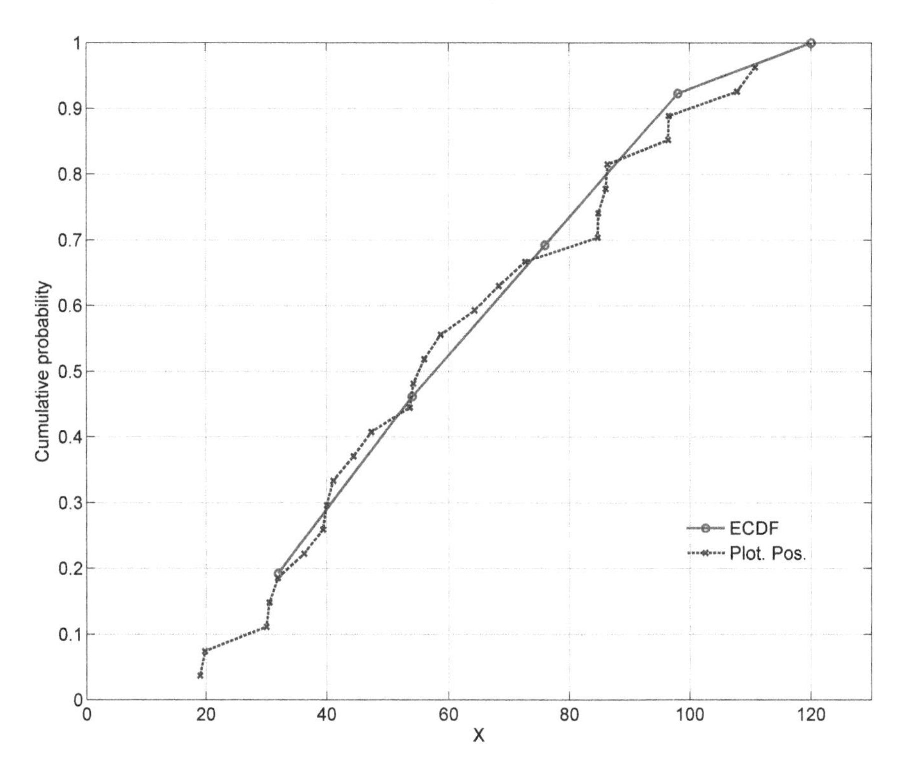

FIGURE 2.4 Cumulative probability of the January precipitation data at Cedars, Quebec, for (a) empirical CDF (solid line with circle marker) and (b) plotting position $i/(n+1)$.

where $\boldsymbol{\mu}$ represents the $p \times 1$ mean value vector of $\boldsymbol{X}$ and where $\boldsymbol{\Sigma}$ is the $p \times p$ variance–covariance matrix of $\boldsymbol{X}$. The symmetric matrix $\boldsymbol{\Sigma}$ must be positive definite (Johnson and Wichern, 2001). This density is simply denoted as $N_p(\boldsymbol{\mu},\boldsymbol{\Sigma})$.

For the case of bivariate normal distribution for $\boldsymbol{X} = [X_1, X_2]^T$, it is explicitly described as

$$f(x_1,x_2) = \frac{1}{2\pi\sigma_1\sigma_2\sqrt{1-\rho^2}} \times \exp\left[-\frac{1}{2(1-\rho^2)}\right.$$
$$\left.\left\{\left(\frac{x_1-\mu_1}{\sigma_1}\right)^2 - 2\rho\left(\frac{x_1-\mu_1}{\sigma_1}\right)\left(\frac{x_2-\mu_2}{\sigma_2}\right) + \left(\frac{x_2-\mu_2}{\sigma_2}\right)^2\right\}\right] \tag{2.52}$$

where μ_1 and σ_1 are the mean and standard deviation of the variable X_1 and ρ is the correlation coefficient which is defined in what follows.

Because $\boldsymbol{Y} = \boldsymbol{y}$, the conditional multivariate normal distribution of $\boldsymbol{X}$ is with mean and variance of

$$\boldsymbol{\mu}_{\mathbf{X}|\mathbf{Y}=\mathbf{y}} = \boldsymbol{\mu}_{\mathbf{X}} + \boldsymbol{\Sigma}_{\mathbf{XY}}\boldsymbol{\Sigma}_{\mathbf{YY}}^{-1}\left(\mathbf{y} - \boldsymbol{\mu}_{\mathbf{Y}}\right) \tag{2.53}$$

$$\Sigma_{X|Y} = \Sigma_{XX} - \Sigma_{XY}\Sigma_{YY}^{-1}\Sigma_{YX} \tag{2.54}$$

Note that the variance–covariance matrix does not depend on the given vector of $\boldsymbol{Y} = \boldsymbol{y}$, though the mean vector does.

For the case of bivariate normal distribution, the conditional distribution is

$$f(x_1 \mid x_2) = \frac{1}{\sqrt{2\pi}\sigma_1\sqrt{1-\rho^2}} \exp\left[-\frac{1}{2}\left(\frac{x_1 - \mu_1 - \rho(\sigma_1/\sigma_2)(x_2 - \mu_2)}{\sigma_1\sqrt{1-\rho^2}}\right)^2\right] \tag{2.55}$$

2.2.2 Covariance and Correlation

The covariance is defined as the central second moment of two variable X_1, X_2, i.e.,

$$\gamma_{12} = Cov(X_1, X_2) = E\left[(X_1 - \mu_1)(X_2 - \mu_2)\right] = E(XY) - E(X)E(Y) \tag{2.56}$$

The correlation coefficient between the two variables is defined as

$$\rho = \frac{Cov(X_1, X_2)}{\sigma_1\sigma_2} \tag{2.57}$$

Note that this correlation coefficient is one of the parameters in the bivariate normal distribution in Eq. (2.52).

From the sample data, the covariance and correlation can be estimated as

$$\hat{\gamma}_{12} = \frac{1}{n}\sum_{i=1}^{n}(x_{1,i} - \bar{x}_1)(x_{2,i} - \bar{x}_2) \tag{2.58}$$

$$\hat{\rho}_{12} = \frac{\sum_{i=1}^{n}(x_{1,i} - \bar{x}_1)(x_{2,i} - \bar{x}_2)}{\sqrt{\sum_{i=1}^{n}(x_{1,i} - \bar{x}_1)^2}\sqrt{\sum_{i=1}^{n}(x_{2,i} - \bar{x}_2)^2}} \tag{2.59}$$

where $\bar{x}_1$ and $\bar{x}_2$ are the sample mean values of the two variables X_1, X_2 as in Eq. (2.36) with the observed data of $x_{1,i}$ and $x_{2,i}$ $i = 1,..,n$.

2.3 RANDOM SIMULATION

2.3.1 Monte Carlo Simulation and Uniform Random Number

Monte Carlo simulation (MCS) is a numerical process that evaluates substantive hypothesis of statistical estimators with a simulation algorithm that draws multiple samples from pseudopopulation. This has been employed as a scientific tool for problems that are analytically intractable and too time consuming for experimentation.

The basic process of MCS is to generate uniform random numbers. A number of approaches have been developed in this respect and linear congruential generators are most commonly employed (Park and Miller, 1988). This approach is to recursively calculate a sequence of integers between 0 and $m - 1$ from a linear transformation as

$$x_i = a + bx_{i-1} \bmod M \tag{2.60}$$

where a and b are increment and multiplier in positive integers as well as M is a large positive integer. Mod indicates the remainder of $(a + bx_{i-1})/M$. When the increment is zero, then it yields the multiplicative congruential generator as

$$x_i = bx_{i-1} \bmod M \tag{2.61}$$

Note that the generated sequence from linear or multiplicative congruential generators ranged from 0 to M and is uniformly distributed denoted as $Unif[0, M]$. If it is divided by M (i.e., x_i/M), it becomes $Unif[0, 1]$ i.e:

$$u_i = \frac{x_i}{M} \sim Unif[0,1] \tag{2.62}$$

The selection of integers such as a, b, and M are highly influential to the generated sequence. However, the generated numbers are repeated with a given period, no matter what integers are selected. Therefore, they are called pseudorandom numbers. Also, a uniform random number generation for a specific range as $u_i^* \sim Unif[s,t]$ can be simply achieved by

$$u_i^* = u_i \times (t - s) + s \tag{2.63}$$

It has been known that it is advantageous to choose M to be a large prime number, such as $2^{31} - 1$ or 2^{32}, which is conjuncted with 32-bit integers for a great speed in generation. Initial seed number (x_0) is chosen randomly among $[0, M - 1]$. Each computer program assigns these integers differently, for example, Borland C/C++ employs $[a, b, m] = [1, 22695477, 2^{32}]$, and Numerical Recipes (Press et al., 2002) suggests $[a, b, m] = [1013804223, 1664525, 2^{32}]$.

Example 2.4

Generate a uniform random number $u_i \sim Unif[0,1]$ with the linear congruential generators and $a = 2$, $b = 1$ and $M = 2^3 - 1 = 7$.

Solution:

Let $k_0 = 3$ and for $i = 1$

$$x_1 = 2 + 1 \times 3 \bmod 7 = rem\left(\frac{5}{7}\right) = 5$$

$$u_1 = \frac{x_1}{M} = \frac{5}{7} = 0.7143$$

The generated sequence is

u = [*0.71, 0.00, 0.29, 0.57, 0.86, 0.14, 0.43*, 0.71, 0.00, 0.29, 0.57, 0.86, 0.14]

It can be noticed that the first seven values are repeated with a period of seven. Highly large number of the integers might downgrade the effect of cyclic behavior of simulated sequences.

2.3.2 Simulation of Probability Distributions

To simulate random numbers from a certain probability distribution, different methods have been proposed, such as decomposition method, rejection method, and inverse transform method. The decomposition method decomposes a complicated probability distribution function into a combination of simpler pdfs, while the rejection method generates a random number from a complex probability distribution by rejecting or accepting a value simulated from a simple distribution with a given probability ratio. The easiest method is the inverse transform method that a generated value is determined with the CDF inverse of a target complex probability with a uniform variate as $u = F^{-1}(x)$ generated from the previous section.

For example, the inverse CDF of the exponential distribution in Eq. (2.33) can be described as

$$x = x_0 + \frac{\alpha}{\beta}\left\{1-[-\ln F]^{\beta}\right\} \tag{2.64}$$

and the CDF of a continuous random variable X has the property of probability integral transform (Kottegoda and Rosso, 1997), such that $u = F_x(X)$ and $u \sim Unif[0,1]$. Therefore,

$$x = x_0 + \frac{\alpha}{\beta}\left\{1-[-\ln u]^{\beta}\right\} \tag{2.65}$$

Note that u can be generated from linear or multiplicative congruential generators as mentioned in the previous section. Also, the inverse of the exponential distribution is shown as in Eq. (2.23) and referred to as

$$x = -\alpha \ln(1-u) \tag{2.66}$$

Example 2.5

Generate a sequence of random number of the exponential distribution in Example 2.1 employing the inverse transform method as well as seven uniform random numbers.

Solution:

The estimated parameter of the exponential distribution (α) in Example 2.1 is $\hat{\alpha} = 60.0423$. The sequence is [0.71 0.00 0.29 0.57 0.86 0.14 0.43]. The sequence of the exponential distribution is simulated from Eq. (2.66) as

$$x = -60.0423 \ln(1-0.71) = 75.22$$

Note that there is a slight difference between the calculated value with 0.71 and the present value with the real value with a further precise decision, which is 0.714285714285714≈0.71 with the rounding to two decimal places. Throughout this book, this difference can happen in some examples. Here, the difference is high due to the effect of log scale in the calculation.

The final result is

$$x = [75.22, 0.00, 20.20, 50.87, 116.84, 9.26, 33.60]$$

2.4 METAHEURISTIC ALGORITHM

Mathematical modeling commonly involves probability distribution models and regression models that contain parameters. The parameters in a model must be estimated with observations. Optimization is a tool to determine the parameter set while meeting a predefined criterion. The optimization for the estimation of parameters is often nonlinear and highly complex so that it should be performed repetitively and numerically, and still the global optimum may not be guaranteed or found. To overcome this drawback, metaheuristic algorithms, such as Genetic Algorithm, Particle Swarm Optimization, and Harmony Search (HS), have been developed. Metaheuristic algorithm is the algorithm that generates a number of possible candidates and evolves a process to reach the best solution. Among others, HS is introduced in this chapter with a simple example. Yoon et al. (2013) suggested that HS is a competitive tool whose search ability outperforms the Genetic Algorithm in the parameter estimation of probability distribution models.

2.4.1 HARMONY SEARCH

The HS metaheuristic algorithm is a phenomenon-mimicking algorithm inspired by the improvisational processes of musicians and was developed by Geem et al. (2001). The algorithm is based on natural musical performance processes in which a musician searches for a better state of harmony. The algorithm is based on mimicking the musical performance that seek a best state (fantastic harmony) in terms of esthetic estimation: the optimization algorithm seeks a best state (global optimum) in terms of objective function evaluation (Geem et al., 2001). To apply HS, the optimization problem is formulated as follows:

Minimize $\boldsymbol{O}_b(\boldsymbol{a})$

subject to $a_i \in \{Low_i, Up_i\}, \quad i = 1,\ldots,N$

where $\boldsymbol{O}_b(\boldsymbol{a})$ is an objective function; $\boldsymbol{a}$ is the set of each decision variable a_i, $i = 1,\ldots, N$, and N is the number of decision variables; and Low_i and Up_i are the upper and lower limits, respectively, of the decision variable a_i.

To solve the optimization problem, the procedure of the HS algorithm is as follows:

1. Generate the harmony memory (HM) containing the initial sets of decision variables randomly up to the HM size (HMS) from a uniform distribution, denoted as

$$a_i^j \sim U[Low_i, Up_i] \tag{2.67}$$

for $i = 1,\ldots,N$; $j = 1,\ldots,$ HMS; $U[a,b]$ represents the uniform distribution ranging from a to b. The generated HM is

$$\text{HM} = \begin{bmatrix} a_1^1 & a_2^1 & \cdots & a_N^1 \\ a_1^2 & a_2^2 & \cdots & a_N^2 \\ & & \vdots & \\ a_1^{HMS} & a_2^{HMS} & \cdots & a_N^{HMS} \end{bmatrix} = \begin{bmatrix} \mathbf{a}^1 \\ \mathbf{a}^2 \\ \vdots \\ \mathbf{a}^{HMS} \end{bmatrix} \tag{2.68}$$

2. Improvise a new set ($\tilde{\mathbf{a}}_i$) by

$$\tilde{a}_i = \begin{cases} \tilde{a}_i \in \left\{a_i^1, a_i^2, \ldots, a_i^{HMS}\right\} & \text{w.p.HMCR} \\ \tilde{a}_i \sim U[Low_i, Up_i] & \text{otherwise} \end{cases} \tag{2.69}$$

where HMCR is the HM considering the rate for all the variables $\tilde{a}_i$, $i = 1,..,N$. This equation implies that each decision variable of a new harmony set is sampled from the same variable of HM in Eq. (2.68) for the probability of HMCR. Otherwise, generate the harmony set from the uniform distribution.
3. Adjust each variable of the improvised new set ($\tilde{\mathbf{a}}_i$) with the probability of the pitch adjusting rate (PAR) as

$$\tilde{a}_i * \begin{cases} = \tilde{a}_i + \varepsilon & \text{w.p.PAR} \\ = \tilde{a}_i & \text{otherwise} \end{cases} \tag{2.70}$$

where ε is generated from $Unif[-h, h]$. Here, h is the arbitrary distance bandwidth. If h is large, the variability of the adjusted value $\tilde{a}_i^*$ is also large.
4. Update the HM by replacing the worst harmony according to the worst objective function with the improvised and adjusted new set.
5. Repeat steps (2)–(4) until the termination criterion is met. The termination criterion commonly employed is the maximum iteration such that if the iteration is reached to the predefined maximum iteration then stop.

The following empirically based HS parameter range is recommended to produce a sufficient solution as 0.7–0.95 for HMCR, 0.2–0.5 for PAR, and 10–50 for HMS (Geem et al., 2001; Geem et al., 2002; Geem, 2006; Geem, 2009). A typical set of the median values such as HMCR = 0.8 and PAR = 0.4 presents a reliable result (Yoon et al., 2013).

Example 2.6

Estimate the parameter (α) of the exponential probability distribution given in Eq. (2.21) for the January precipitation data at Cedars, Quebec, in Table 2.1 with HMS by maximizing the LL.

Solution:

The objective function is defined by the negative LL as

$$O_b(\alpha) = -LL(x_1, x_2, \ldots, x_n) = -\sum_{i=1}^{n} \log\left\{\frac{1}{\alpha}\exp\left(-\frac{x}{\alpha}\right)\right\} = \sum_{i=1}^{n}\left(\log\alpha + \frac{x}{\alpha}\right)$$

The negative of *LL* changes the maximization to minimization of the objective function. The range of parameter is set to be [$\alpha_{Low} = 0.001$, $\alpha_{Up} = 1000$]. The other setting is HMS = 10, Maximum iteration = 1,000,000, HMCR = 0.8, PAR = 0.4, b = (Up-Low)/10,000 = 0.1. Note that HMCR and PAR values are assigned at the median of the suggested range from the literature as Yoon et al. (2013).

1. Generate the HM from the uniform distribution between [$\alpha_{Low} = 0.001$, $\alpha_{Up} = 1{,}000$] as $\alpha \sim Unif[\alpha_{Low}, \alpha_{Up}]$. A uniform random number generation with the lower and upper range can be made as in Eq. (2.63). Let the generated HM from the uniform random number be assumed as

$$\text{HM} = \begin{bmatrix} 988.583 \\ 442.467 \\ 324.092 \\ 379.219 \\ 694.809 \\ 235.772 \\ 464.076 \\ 154.830 \\ 968.887 \\ 31.516 \end{bmatrix} \quad \text{O}_b = \begin{bmatrix} 180.882 \\ 161.930 \\ 155.124 \\ 158.508 \\ 172.381 \\ 148.656 \\ 163.005 \\ 141.183 \\ 180.391 \\ 139.247 \end{bmatrix}$$

Here, the objective function (O_b) as given earlier needs to be estimated with the candidate HM set. For example, for $\alpha_1 = 988.583$ for the precipitation data in the second column of Table 2.1,

$$O_b = \left(\log 988.583 + \frac{39.4}{988.583}\right) + \cdots \left(\log 988.583 + \frac{68.3}{988.583}\right) = 180.882$$

2. Improvise a new parameter set, which means creating a new parameter set with the probability of HMCR (here HMCR = 0.8 as previously assigned). Note that a parameter set is just one parameter of α in the current example.

It is mathematically expressed as

$$\alpha_{new} = \begin{cases} \alpha \sim Unif[\alpha_{Low}, \alpha_{Up}] & \text{w.p. HMCR}(=0.8) \\ \alpha & \text{from step(3)} \end{cases}$$

In practice, generate a uniform random number ($\delta_c \sim Unif[0,1]$). If $\delta_c \le 0.8$ (for example, $\delta_c = 0.632$ is generated from the uniform distribution [0, 1]), perform step (3). If $\delta_c > 0.8$ (for example, $\delta_c = 0.895$ is generated), generate one new parameter set as in step (1) to get a new parameter set (here, α_{new}).

3. Select a parameter set among HM with equal probability for each parameter set. Assuming the seventh value (464.076) selected, adjust the parameter with the probability of PAR (here, PAR = 0.4). If a uniform random number $\delta_p \le$ PAR (e.g., $\delta_p = 0.213$), then

$$\alpha_{new} = \alpha_{select} + \delta_{PK} \times b = 464.076 + (0.490) \times 0.1 = 464.125$$

Here, δ_{PK} is the uniform random number between [–1,1], assuming that 0.490 is generated. The objective function of α_{new} is $O_b = 163.008$.

4. Update the worst performed parameter set according to its objective function if the new parameter set (here, α_{new}) has the better objective function. The worst parameter is 988.583 with $O_b = 180.882$, and it is replaced with $\alpha_{new} = 464.125$ and $O_b = 163.008$. The HM is updated as

$$\text{HM} = \begin{bmatrix} 464.125 \\ 442.467 \\ 324.092 \\ 379.219 \\ 694.809 \\ 235.772 \\ 464.076 \\ 154.830 \\ 968.887 \\ 31.516 \end{bmatrix} \qquad \text{O}_b = \begin{bmatrix} 163.008 \\ 161.930 \\ 155.124 \\ 158.508 \\ 172.381 \\ 148.656 \\ 163.005 \\ 141.183 \\ 180.391 \\ 139.247 \end{bmatrix}$$

5. Continue the steps (2)–(4) until the iteration reaches the maximum iteration of 10,000. The final HM and O_b become

$$\text{HM} = \begin{bmatrix} 60.0423 \\ 60.0423 \\ 60.0423 \\ 60.0423 \\ 60.0423 \\ 60.0423 \\ 60.0423 \\ 60.0423 \\ 60.0423 \\ 60.0423 \end{bmatrix} \quad \text{O}_b = \begin{bmatrix} 132.4713 \\ 132.4713 \\ 132.4713 \\ 132.4713 \\ 132.4713 \\ 132.4713 \\ 132.4713 \\ 132.4713 \\ 132.4713 \\ 132.4713 \end{bmatrix}$$

The final parameter α_{final} is 60.0423.

2.5 SUMMARY AND CONCLUSION

In statistical downscaling, it is critical to have a statistical background in analyzing the outputs as well as downscaling itself. The statistical background that is used in this book is explained in this chapter. Further advanced statistical techniques are also explained in each chapter.

3 Data and Format Description

3.1 GCM DATA

The recent study of the Intergovermental Panel on Climate Change (IPCC) Fifth Assessment Report (AR5) through Coupled Model Intercomparison Project, employs a number of global climate models (GCMs) and Earth system Models of Intermediate Complexity (EMICs). The characteristics and evaluation of the models employed are described in the AR5 report (Chapter 9 of Flato et al. (2013)). Almost all the GCM data employed are available at http://pcmdi-cmip.llnl.gov/cmip5/availability.html linked to the following four websites:

- Program for Climate Model Diagnosis and Intercomparison (PCMDI): http://pcmdi9.llnl.gov/
- British Atmospheric Data Centre (BADC): http://esgf-index1.ceda.ac.uk
- Deutsches Klimarechenzentrum (DKRZ) –German Climate Computing Center: http://esgf-data.dkrz.de
- National Computational Infrastructure (NCI): http://esg2.nci.org.au

Since the focus of this book is on statistical downscaling, a full version of climate models (i.e., GCMs) is commonly employed. The GCMs employed in the IPCC AR5 report are presented in Table 3.1 along with their horizontal grid. Figure 3.1 presents an example of data acquisition from the first website PCMDI, searching for RCP4.5, precipitation, and HadGEM scenario. One can select more detailed options from the left panel of this website. Here, HadGEM2-AO is selected from the Model option in Figure 3.1. If "Show Files" is clicked for the second one, it is displayed as shown in Figure 3.1. The fifth one is selected among the listed files by clicking HTTP Download. Data Access Login website is shown and the Login ID is obtained by sign in.

For example, the name of the downloaded file from the website is

pr_day_HadGEM2-AO_rcp45_r1i1p1_20060101–21001230.nc

Here, "pr" indicates the name of the variable (precipitation); "day" is the time frequency (daily); HadGem2-AO is the GCM model name; rcp45 is the climate scenario name; r1i1p1 is the ensemble name, and 20060101–21001230 is the contained time length.

The format of downloaded data is mostly "nc"(i.e., NetCDF format file). This NetCDF format is an open international standard of the Open Geospatial Consortium developed by the University Corporation for Atmospheric Research. The extraction of the NetCDF format is available through various programs, such as C, C++, Fortran, R, Matlab, and Java (www.unidata.ucar.edu/software/netcdf/).

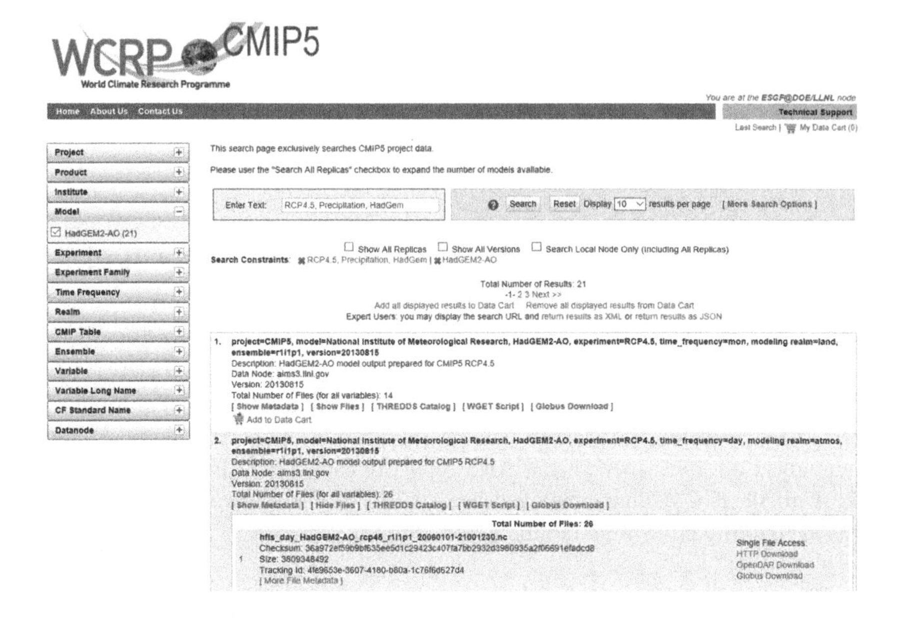

FIGURE 3.1 Data acquisition of GCM at https://esgf-node.llnl.gov/search/cmip5/ for HadGEM2-AO, RCP4.5 daily sequence.

3.2 REANALYSIS DATA

Climate reanalysis numerically describes the past climate by combining models with observations. The analysis estimates atmospheric parameters (e.g., air temperature, pressure, and wind) and surface parameters (e.g., rainfall, soil moisture content, and sea-surface temperature). These estimates are globally produced and span a long time period that can extend back by decades or more. These estimates clearly show past weather conditions, while various instrumental observations are combined.

This reanalysis data is critical to statistical downscaling, since statistical downscaling employs the statistical relationship between observations and climatic variables in GCM outputs. A regular GCM output does not simulate the real past conditions of atmospheric variables but just virtual conditions according to the restrictions of the GCM parameters, such as doubling CO_2. Therefore, in statistical downscaling, the relationship between a target local variable and GCM output variables can be established through the observation of a local variable and the reanalysis data of atmospheric variables. Then, GCM outputs are used to predict future conditions of the local variable by replacing the reanalysis data with GCM outputs. Reanalysis datasets available are described in Table 3.2.

3.3 RCM DATA

While GCMs are widely used in the assessment of climate change impact, their horizontal resolutions are too coarse to assess the climate change impact in a local scale. To bridge this gap, regional climate models (RCMs) have been a useful tool forced

TABLE 3.1
GCMs Employed in the Study of IPCC AR5

Modeling Center	Model	Horizontal Grid	Institution
BCC	BCC-CSM1.1 BCC-CSM1.1(m)	T42 T42L26 T106	Beijing Climate Center, China Meteorological Administration
CCCma	CanCM4 CanESM2	Spectral T63 Spectral T63	Canadian Centre for Climate Modeling and Analysis
CMCC	CMCC-CESM CMCC-CM CMCC-CMS	3.75°×3.75° 0.75°×0.75° 1.875°×1.875°	Centro Euro-Mediterraneo per I Cambiamenti Climatici
CNRM-CERFACS	CNRM-CM5	1.4° TL127	Centre National de Recherches Meteorologiques/Centre Europeen de Recherche et Formation Avancees en Calcul Scientifique
COLA and NCEP	NCEP-CFSv2	.9375 T126	Center for Ocean-Land-Atmosphere Studies and National Centers for Environmental Prediction
CSIRO-BOM	ACCESS1.0 ACCESS1.3	192×145 N96 192×145 N96	CSIRO (Commonwealth Scientific and Industrial Research Organisation, Australia), and BOM (Bureau of Meteorology, Australia)
CSIRO-QCCCE	CSIRO-Mk3.6.0	1.875°×1.875°	Commonwealth Scientific and Industrial Research Organisation in collaboration with the Queensland Climate Change Centre of Excellence
EC-EARTH	EC-EARTH	1.1255°longitudinal spacing, Gaussian grid T159L62	EC-EARTH consortium
FIO	FIO-ESM	T42[a]	The First Institute of Oceanography, SOA, China
GCESS	BNU-ESM	T42	College of Global Change and Earth System Science, Beijing Normal University
IPSL	IPSL-CM5A-LR	96×95 equivalent to 1.9°×3.75° LMDZ96×95	Institut Pierre-Simon Laplace
	IPSL-CM5A-MR	144×143 equivalent to 1.25°×2.5° LMDZ144×143	
	IPSL-CM5B-LR	96×95 equivalent to 1.9°×3.75° LMDZ96×95	

(*Continued*)

TABLE 3.1 (*Continued*)
GCMs Employed in the Study of IPCC AR5

Modeling Center	Model	Horizontal Grid	Institution
LASG-CESS	FGOALS-g2	2.8125°×2.8125°	LASG, Institute of Atmospheric Physics, Chinese Academy of Sciences; and CESS, Tsinghua University
LASG-IAP	FGOALS-s2	R42(2.81°×1.66°)	LASG, Institute of Atmospheric Physics, Chinese Academy of Sciences; and CESS, Tsinghua University
MIROC	MIROC4h	0.5625°×0.5625° T213	Atmosphere and Ocean Research Institute (The University of Tokyo), National Institute for Environmental Studies, and Japan Agency for Marine-Earth Science and Technology
	MIROC5	1.40625°×1.40625° T85	
MIROC	MIROC-ESM	2.8125°×2.8125° T42	Japan Agency for Marine-Earth Science and Technology, Atmosphere and Ocean Research Institute (The University of Tokyo), and National Institute for Environmental Studies
	MIROC-ESM-CHEM	2.8125°×2.8125° T42	
MOHC (additional realizations by INPE)	HadCM3	N48L19 3.75°×2.5°	Met Office Hadley Centre (additional HadGEM2-ES realizations contributed by Instituto Nacional de Pesquisas Espaciais)
	HadGEM2-A	1.875° in longitude by 1.25° in latitude N96	
	HadGEM2-CC	1.875° in longitude by 1.25°in latitude N96	
	HadGEM2-ES	1.875° in longitude by 1.25° in latitude N96	
MPI-M	MPI-ESM-LR	approx. 1.8° T63	Max Planck Institute for Meteorology (MPI-M)
	MPI-ESM-MR	approx. 1.8° T63	
	MPI-ESM-P	approx. 1.8° T63	
MRI	MRI-AGCM3.2H	640×320 TL319	Meteorological Research Institute
	MRI-AGCM3.2S	1920×960 TL959	
	MRI-CGCM3	320×160 TL159	
	MRI-ESM1	320×160 TL159	
NASA GISS	GISS-E2-H	2° latitude × 2.5°longitude F	NASA Goddard Institute for Space Studies
	GISS-E2-H-CC	Nominally 1	
	GISS-E2-R	2° latitude×2.5° longitude F	
	GISS-E2-R-CC	Nominally 1°	

(*Continued*)

TABLE 3.1 (*Continued*)
GCMs Employed in the Study of IPCC AR5

Modeling Center	Model	Horizontal Grid	Institution
NCAR	CCSM4	0.9°×1.25°	National Center for Atmospheric Research
NCC	NorESM1-M	Finite Volume 1.9° latitude, 2.5° longitude	Norwegian Climate Centre
	NorESM1-ME	Finite Volume 1.9° latitude, 2.5° longitude	
NOAA GFDL	GFDL-CM2.1	2.5° longitude, 2° latitude M45L24	Geophysical Fluid Dynamics Laboratory
	GFDL-CM3	~200 km C48L48	
	GFDL-ESM2G	2.5° longitude, 2° latitude M45L24	
	GFDL-ESM2M	2.5° longitude, 2° latitude M45L24	
	GFDL-HIRAM-C180	Averaged cell size: approximately 50×50 km. C180L32	
	GFDL-HIRAM-C360	Averaged cell size: approximately 25×25 km. C360L32	
NSF-DOE-NCAR	CESM1(BGC)	0.9°×1.25°	National Science Foundation, Department of Energy, National Center for Atmospheric Research
	CESM1(CAM5)	0.9°×1.25°	
	CESM1(FASTCHEM)	0.9°×1.25°	
	CESM1(WACCM)	1.9°×1.25°	

[a] T42 indicates spectral resolution with 64×128 for the numbers of latitude and longitude, respectively, around 5.61 at the equator.

by time-dependent meteorological lateral boundary conditions (Feser et al., 2011). The Coordinate Regional Climate Downscaling Experiment (CORDEX) separates the globe into domains, such as South America, Central America, North America, Europe, Africa, South Asia, East Asia, Central Asia, Australasia, Antarctica, Arctic, Mediterranean, and South East Asia, for which regional downscaling takes place. It provides a number of RCM outputs and related information at (www.cordex.org/output/cordex-rcm-list.html). The following websites allow to download the output of RCMs:

- CORDEX East Asia: https://cordex-ea.climate.go.kr/
- CORDEX North America: https://na-cordex.org/
- CORDEX Europe: www.euro-cordex.net/

Note that statistical downscaling should also be applied to RCM outputs when site-specific application is required, and the temporal scale of the outputs (e.g., hourly) is still not enough for small watersheds (e.g., urban areas).

TABLE 3.2
Reanalysis Data Quality, Proven by IPCC[a]

Data Name	Time Span	Spatial Res.	Website
NCEP Reanalysis Data	1948–present	2.5°	www.esrl.noaa.gov/psd/data/gridded/data.ncep.reanalysis.html
ECMWF 40-year Reanalysis (ECMWF ERA-40)	1958–2001	125 km	http://apps.ecmwf.int/datasets/data/era40-daily/levtype=sfc/
ECMWF Interim Reanalysis (ECMWF ERA-Int)	1979–present	1.5°	www.ecmwf.int/en/research/climate-reanalysis/era-interim
Japanese 55-year Reanalysis (JRA-55)	1958–2012	1.25°	http://jra.kishou.go.jp/JRA-55/index_en.html
Modern Era Retrospective-analysis for Research and Applications (MERRA-2)	1980–present	50 km	https://gmao.gsfc.nasa.gov/reanalysis/MERRA-2/
National Centers for Environmental Prediction (NCEP) and Climate Forecast System Reanalysis (CSFR) data (NCEP-CFSR)	1979–2011	0.5°	http://cfs.ncep.noaa.gov/cfsr/
NCEP-DOE Reanalysis 2 project (NCEP-DOE R2)	1979–present	1.875°	www.esrl.noaa.gov/psd/data/gridded/data.ncep.reanalysis2.html
NOAA CIRES Twentieth Century Global Reanalysis Version (NOAA_CIRES20thC_ReaV2)	1851–2012	2.0°	www.esrl.noaa.gov/psd/data/gridded/data.20thC_ReanV2c.html

[a] Website: www.ipcc-data.org/guidelines/pages/approvedDatasetLinks.html.

3.4 SUMMARY AND CONCLUSION

Since statistical downscaling employs statistical relationships between observed local variables and climate variables, acquisition of appropriate datasets is important. The brief explanation of climate model outputs is presented in this chapter, including RCM output. A number of GCM and RCM outputs are available on websites. Explanations of observed data are omitted in this chapter, since local observed dataset is too large and specific to cover.

4 Bias Correction

4.1 WHY BIAS CORRECTION?

Global climate model (GCM) and regional climate model (RCM) outputs reportedly exhibit systematic biases relative to observations from different sources, such as errors in convection parameterization and unresolved fine-scale orography, unrealistic large-scale variability, and unpredictable internal variability different from observations (Cannon et al., 2015). The outputs simulated from the models are statistically different with respect to observations. To perform statistical downscaling, this bias must be corrected. A number of bias correction methods have been developed for adjusting mean, variance, and higher moment of distributions as well as all quantiles. For adjusting the moments of model outputs to the ones of observations, the delta method has been applied, while quantile mapping (QM) has been used for all quantiles. Note that GCM outputs are for two separate periods, such as base and future. A base period indicates a simulation length that considers observational climate conditions, while a future period presents the simulation length that takes into account climate forcings. Therefore, future GCM outputs are generally targeted data for bias correction with the comparison between the statistics of the base GCM outputs and observations.

Let us assume that the target model output to adjust is the future-projected model value, denoted as $x_{m,fut}$ with the observed data y_{obs}. The bias correction is the procedure that transforms (or maps) the target model value $x_{m,fut}$ into the observation domain using the relationship with the observations and the model value for the base period, $x_{m,base}$. The final bias-corrected future model output is denoted as $\hat{y}_{m,fut}$. Note that the base period and the observational period are normally the same, but not necessarily.

4.2 OCCURRENCE ADJUSTMENT FOR PRECIPITATION DATA

It has been known that precipitation values from GCMs or RCMs are more frequently generated than observations. To remove this difference in the occurrence of precipitation, some fraction of model precipitation values must be treated to zero values. This can be done by comparing the number of observed and model precipitation values that are greater than zero.

Example 4.1

Adjust the occurrence of the precipitation dataset in the third column of Table 4.1.

The second column presents the observed precipitation data while the third column presents the model simulated data. The number of precipitation values that are greater than zero is 19 and 37 among 40 values for the observed and simulated data, respectively.

TABLE 4.1
Occurrence Adjustment for Precipitation Data (Obs = observation)

	Obs	GCMbase	Obs[a] Sort	GCMbase[b] Sort	GCMbase Adjust
1	0.00	0.73	115.00	131.46	0.00
2	0.00	1.69	95.50	51.11	0.00
3	95.50	0.01	55.00	49.93	0.00
4	37.00	2.52	40.00	41.27	2.52
5	0.00	49.93	37.00	27.56	49.93
6	13.50	41.27	26.50	21.84	41.27
7	1.00	2.48	17.50	19.27	0.00
8	55.00	0.14	15.50	16.40	0.00
9	1.00	3.73	13.50	15.24	3.73
10	17.50	1.48	11.00	14.85	0.00
11	0.00	15.24	8.50	11.71	15.24
12	0.00	27.56	3.00	9.41	27.56
13	0.00	19.27	1.50	5.71	19.27
14	8.50	131.46	1.00	5.39	131.46
15	11.00	21.84	1.00	5.16	21.84
16	40.00	11.71	1.00	3.73	11.71
17	0.00	16.40	0.50	2.52	16.40
18	15.50	5.39	0.20	2.48	5.39
19	0.00	1.43	0.20	2.44	0.00
20	1.50	1.19	0.00	2.33	0.00
21	0.00	0.08	0.00	2.18	0.00
22	0.00	0.07	0.00	1.69	0.00
23	0.00	2.33	0.00	1.67	0.00
24	0.20	14.85	0.00	1.52	14.85
25	0.00	0.34	0.00	1.48	0.00
26	0.00	51.11	0.00	1.43	51.11
27	0.00	1.67	0.00	1.19	0.00
28	0.00	0.10	0.00	0.73	0.00
29	0.00	0.28	0.00	0.54	0.00
30	0.00	0.50	0.00	0.50	0.00
31	3.00	5.71	0.00	0.34	5.71
32	0.00	9.41	0.00	0.28	9.41
33	0.00	5.16	0.00	0.14	5.16
34	0.20	1.52	0.00	0.10	0.00
35	115.00	2.18	0.00	0.08	0.00
36	0.00	0.54	0.00	0.07	0.00
37	0.00	2.44	0.00	0.01	0.00
38	1.00	0.00	0.00	0.00	0.00
39	26.50	0.00	0.00	0.00	0.00
40	0.50	0.00	0.00	0.00	0.00
		$N(x > 0)$	19	37	19

[a] Observation data in the second column were sorted in descending order.
[b] GCM data in the third column were sorted in descending order.

Solution:

To find the threshold for the model data whether to set zero, both observation and model data are sorted in a descending order then, let the GCM simulation data have the same number of occurrence values as the observation data with the way that: the model data that is smaller than 2.44 mm (as shown in the fifth column of Table 4.1) is set to be zero as in the sixth column.

In case the number of observations and model data are different, it can be done by estimating the occurrence probability for the observed data and applying the probability to the model data. The occurrence probability of the observed data is 19/40 = 0.475 (i.e., n_1/N, here N is the number of the data and n_1 is the number of the data that are greater than zero.). The 52.5 (=100%–47.5%) percentile of the model data can be the threshold to decide whether the data should be zero or not. In this example, the 52.5 percentile of the model value is 2.44 mm. Therefore, the model precipitation values that are less than 2.4 mm should be set to zero.

This procedure is also physically feasible, since a small amount of rain is considered as no rain in a real observation.

4.3 EMPIRICAL ADJUSTMENT (DELTA METHOD)

For precipitation, a future model value ($x_{m,fut}$) is standardized at first as

$$z_{m,fut} = \frac{x_{m,fut} - \mu_{m,fut}}{\sigma_{m,fut}} \tag{4.1}$$

where $\mu_{m,fut}$ and $\sigma_{m,fut}$ are the mean and standard deviation of the future model values, respectively. Then, the adjusted mean and standard deviation are applied.

$$\hat{y}_{m,fut} = z_{m,fut}\tilde{\sigma}_{m,fut} + \tilde{\mu}_{m,fut} \tag{4.2}$$

where

$$\tilde{\sigma}_{m,fut} = \frac{\sigma_{obs}}{\sigma_{m,base}}\sigma_{m,fut} \tag{4.3}$$

and

$$\tilde{\mu}_{m,fut} = \frac{\mu_{obs}}{\mu_{m,base}}\mu_{m,fut} \tag{4.4}$$

or

$$\tilde{\mu}_{m,fut} = \mu_{m,fut} + \left(\mu_{obs} - \mu_{m,base}\right) \tag{4.5}$$

Here, the adjusted mean with Eq. (4.4) is employed for precipitation, while the adjustment in Eq. (4.5) is applied for temperature.

To consider the seasonal cycles of precipitation and temperature, the monthly or seasonal mean and standard deviation are applied to each month or season. In other words,

$$\hat{y}^j_{m,fut} = z^j_{m,fut}\tilde{\sigma}^j_{m,fut} + \tilde{\mu}^j_{m,fut} \tag{4.6}$$

where j denotes a season or month. For example, if daily precipitation for month j is the target model output, the mean and standard deviation of daily precipitation are estimated only with the dataset for the jth month.

Example 4.2

Correct the bias in the future GCM precipitation data in the fourth column of Table 4.2 employing the delta method, especially with Eqs. (4.4) and (4.3).

Solution:

The observation data (y_{obs}) and the GCM output for the base period ($x_{m,base}$) are presented in the second and third columns of Table 4.2, respectively, while the

TABLE 4.2
Example of Empirical Adjustment for Selected Daily Precipitation (mm)

	Obs.	GCMbase, x_{base}	GCMfut, x_{fut}	Standardized	BCGCMfut, $\hat{y}_{fut}$
1	55.00	1.639	3.739	–0.068	11.336
2	1.00	2.387	1.741	–0.376	7.940
3	17.50	0.009	0.380	–0.586	5.627
4	8.50	0.010	3.139	–0.160	10.316
5	11.00	10.709	17.031	1.982	33.927
6	40.00	76.989	5.839	0.256	14.905
7	15.50	21.872	0.779	–0.524	6.306
8	1.50	4.121	3.806	–0.058	11.450
9	0.20	10.588	13.799	1.484	28.434
10	3.00	0.667	0.323	–0.595	5.530
11	0.20	0.047	1.044	–0.484	6.756
12	115.00	0.064	0.104	–0.629	5.158
13	1.00	0.067	2.540	–0.253	9.298
14	26.50	0.085	14.144	1.537	29.020
15	0.50	0.457	0.056	–0.636	5.076
16	19.00	0.285	0.074	–0.633	5.107
17	0.50	15.233	0.033	–0.640	5.037
18	16.00	19.541	11.273	1.094	24.141
19	15.00	1.925	23.524	2.984	44.961
20	57.50	0.023	0.472	–0.572	5.784
21	25.00	0.011	0.403	–0.582	5.667
22	1.00	1.685	0.047	–0.637	5.062
23	0.50	0.060	0.038	–0.639	5.047
24	0.50	1.202	0.137	–0.624	5.213
25	4.50	0.023	0.013	–0.643	5.004
26	8.00	8.656			
27	6.50	1.366			
28	0.50	1.718			
29	39.50	5.042			
30	38.50	5.157			
31	25.50	0.094			
Mean	17.884	6.185	4.179	$\tilde{\mu}_{m,fut}$	12.084
Variance	599.509	207.545	42.025	$\tilde{\sigma}_{m,fut}$	11.018
Std	24.087	14.172	6.483		

target future GCM is in the sixth column. The empirical adjustment for the first model output ($x_{m,fut}$ = 3.738) can be bias-corrected with the following procedure.

1. The future GCM data is standardized as

$$z_{m,fut} = \frac{x_{m,fut} - \mu_{m,fut}}{\sigma_{m,fut}} = \frac{3.739 - 4.179}{\sqrt{42.025}} = -0.068$$

All the standardized future GCM data is shown in the fifth column of Table 4.2.

2. The adjusted mean and standard deviation from Eqs. (4.4) and (4.3) are estimated as

$$\tilde{\mu}_{m,fut} = \frac{\mu_{obs}}{\mu_{m,base}} \mu_{m,fut} = \frac{17.884}{6.185} \times 4.179 = 12.084$$

$$\tilde{\sigma}_{m,fut} = \frac{\sigma_{obs}}{\sigma_{m,base}} \sigma_{m,fut} = \frac{\sqrt{599.509}}{\sqrt{207.545}} \sqrt{42.025} = 11.018$$

3. The adjusted mean and standard deviation estimated from the earlier equation are applied as

$$\hat{y}_{m,fut} = z_{m,fut} \tilde{\sigma}_{m,fut} + \tilde{\mu}_{m,fut} = -0.068 \times 11.018 + 12.084 = 11.336$$

All the adjusted GCM future precipitation are presented in the sixth column of Table 4.2.

4.4 QUANTILE MAPPING

4.4.1 General Quantile Mapping

QM allows the probability distribution of the model outputs from GCMs and RCMs (x) to be adjusted to the probability distribution of the observed data (y) by matching the cumulative distribution function (CDF, $F(x;\boldsymbol{\theta})$, where $\boldsymbol{\theta}$ represents the parameter set) values of the two distributions. Through QM, the CDF of the RCM output data is transferred to the observed data. The traditional QM can be defined by

$$\hat{y} = F_o^{-1}(F_{m,base}(x)) \tag{4.7}$$

where F_o^{-1} represents an inverse function of CDF for the observed data and $F_{m,base}$ is the CDF of the model output from RCM or GCM fitted to the GCM outputs for the base period.

Different CDFs for QM, such as Gamma, Exponential, Double Exponential as well as mixture distributions, have been applied in the literature.

Example 4.3

Correct the bias of the future GCM precipitation in the seventh column of Table 4.3 employing QM with the exponential and Gamma distributions.

Solution:

The observations and the GCM data are presented in the second and third columns of Table 4.3. These are precipitation datasets that are greater than zero

TABLE 4.3
Example of QM with Gamma Distribution for Selected Daily Precipitation (mm)

	Obs.	GCMbase, x_{base}	F_o	$F_{m,base}$	BCGCMbase, $\hat{y}_{base}$	GCMfut, x_{fut}	BCGCMfut, $\hat{y}_{fut}$
1	55.00	1.639	0.923	0.617	14.040	3.739	20.454
2	1.00	2.387	0.171	0.659	16.605	1.741	14.417
3	17.50	0.009	0.672	0.238	1.888	0.380	7.638
4	8.50	0.010	0.497	0.243	1.961	3.139	18.836
5	11.00	10.709	0.557	0.838	34.867	17.031	45.231
6	40.00	76.989	0.867	0.992	121.227	5.839	25.422
7	15.50	21.872	0.641	0.916	52.443	0.779	10.226
8	1.50	4.121	0.211	0.723	21.428	3.806	20.628
9	0.20	10.588	0.073	0.837	34.653	13.799	40.107
10	3.00	0.667	0.301	0.525	9.589	0.323	7.163
11	0.20	0.047	0.073	0.323	3.441	1.044	11.561
12	115.00	0.064	0.990	0.342	3.858	0.104	4.624
13	1.00	0.067	0.171	0.345	3.925	2.540	17.082
14	26.50	0.085	0.775	0.360	4.290	14.144	40.669
15	0.50	0.457	0.119	0.490	8.224	0.056	3.668
16	19.00	0.285	0.693	0.450	6.818	0.074	4.077
17	0.50	15.233	0.119	0.878	42.418	0.033	3.009
18	16.00	19.541	0.649	0.905	49.028	11.273	35.855
19	15.00	1.925	0.633	0.634	15.074	23.524	54.809
20	57.50	0.023	0.930	0.283	2.649	0.472	8.335
21	25.00	0.011	0.761	0.247	2.029	0.403	7.823
22	1.00	1.685	0.171	0.620	14.212	0.047	3.450
23	0.50	0.060	0.119	0.338	3.767	0.038	3.196
24	0.50	1.202	0.119	0.584	12.274	0.137	5.132
25	4.50	0.023	0.369	0.283	2.649	0.013	2.178
26	8.00	8.656	0.484	0.813	31.118		
27	6.50	1.366	0.440	0.597	12.969		
28	0.50	1.718	0.119	0.622	14.334		
29	39.50	5.042	0.864	0.747	23.636		
30	38.50	5.157	0.859	0.750	23.900		
31	25.50	0.094	0.766	0.367	4.455		
Mean	17.884	6.185			19.644	4.179	
Var.	599.509	207.545			559.314	42.0255	

(i.e., only nonzero precipitation values are taken from daily precipitation). Note that the number of data from observations and the GCM (or RCM) can often be different.

Here, the same number of data is presented just for convenience. The mean and variance of the observations and the GCM data are estimated and shown in the last two rows of Table 4.3.

1. Exponential distribution: F_m and F_o fit to an exponential distribution in (2.21). The parameters of the exponential distribution for observations and GCM data are 17.63(α_o) and 6.39(α_m), respectively, since the parameter of the exponential distribution is the same as a mean value shown in Eq. (2.24). The CDF can be estimated as

$$\hat{y} = F_o^{-1}(F_{m,base}(x;\alpha_m);\alpha_o) = -\alpha_o \ln\{1 - F_{m,base}(x;\alpha_m)\}$$

$$= -\alpha_o \ln\left[1 - 1 + \exp\left(-\frac{x}{\alpha_m}\right)\right] = \frac{\alpha_o}{\alpha_m}x$$

For, $x_{fut}^1 = 3.739$

$$\hat{y} = 17.63 / 6.39 \times 3.739 = 10.32$$

2. Gamma distribution: F_m and F_o fit a gamma distribution given by Eq. (2.26). The parameters of the gamma distribution for both datasets are estimated with the method of moments using Eqs. (2.30) and (2.29). For example, parameters α and β for the observations are

$$\hat{\beta}_o = \frac{\hat{\sigma}^2}{\hat{\mu}} = \frac{599.51}{17.884} = 33.522$$

$$\hat{\alpha}_o = \frac{\hat{\mu}}{\hat{\beta}} = \frac{17.884}{33.522} = 0.534$$

The estimated parameters are as follows:

$$\hat{\alpha}_o = 0.534 \text{ and } \hat{\alpha}_{m,base} = 0.183$$

$$\hat{\beta}_o = 33.522 \text{ and } \hat{\beta}_{m,base} = 33.557$$

From the estimated parameters, QM can be done by applying Eq. (4.7). For example, $x_{fut}^1 = 3.739$

$$\hat{y} = F_o^{-1}\left(F_{m,base}(3.739)\right) = F_o^{-1}(0.711) = 20.454$$

where $F(x;\alpha,\beta)$ defines the CDF with parameters α and β. Figure 4.1 presents the related datasets and their corresponding CDFs. The empirical CDF (ECDF) is estimated with the general estimation rule, i.e., ECDF

$$\hat{F}\left(x_{(i)}\right) = \frac{i}{(N+1)} \tag{4.8}$$

where $x_{(i)}$ is the ith increasing-ordered value, and N is the number of data.

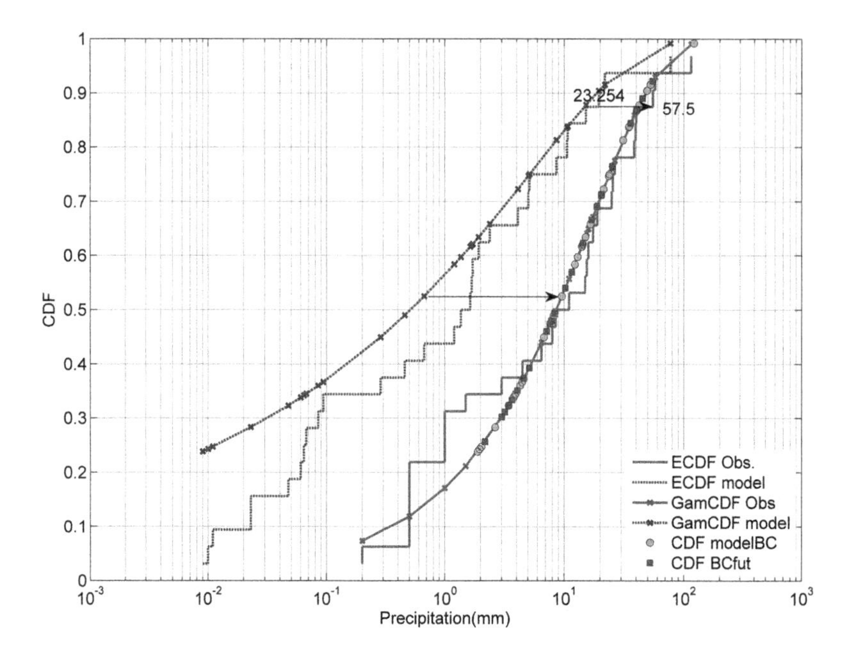

FIGURE 4.1 **(See color insert.)** QM with gamma and empirical distributions (see Table 4.3). The Gamma CDFs for observed and GCM data are presented with smooth red solid and blue dotted lines with cross makers. The specific values for the bias-corrected data for the observed period (CDF model BC) and future period (CDF BCfut) are shown with circle and square markers. In addition, the ECDFs of the observed and GCM model data are shown with the stairs as red solid and blue dotted lines, respectively.

4.4.2 Nonparametric Quantile Mapping

The nonparametric QM can be done by employing ECDF in Eq. (4.8) without any assumption of a parametric distribution for observations or model data.

Example 4.4

Employing the dataset in Table 4.3, perform nonparametric QM

Its procedure is presented in Table 4.4 and is explained as follows:

1. Sort the observations and the model output of the base (or observational) period with an increasing order as shown in the second and third columns of Table 4.4, respectively.
2. Estimate the corresponding ECDFs of the observations and the model output with Eq. (4.8) shown in the fourth and fifth columns, respectively. For example, the observation CDF of the fourth order $F_o(4) = 4/32 = 0.125$.
3. Estimate the quantile of the observational ECDF (i.e., $F_o^{-1}(F_{m,base})$) as follows: (a) Find the nearest value among the model output of the base period to the target future model output (e.g., x_{fut} = 23.524 and x_{base}(nearest) = 21.872; and (b) assign its corresponding ECDF in the fifth column for the ECDF of $F_{m,base}$(here $F_{m,base}$ = 0.938); and (3); and (c) assign its corresponding

TABLE 4.4
Example of QM with Empirical Distribution with the Dataset in Table 4.3

	Obs. ord[a]	GCMbase ord[b] x_{base}	F_o	$F_{m,base}$	GCMfut, x_{fut}	BCGCM fut, $\hat{y}_{fut}$
1	0.20	0.009	0.031	0.031	3.739	19.000
2	0.20	0.010	0.063	0.063	1.741	15.500
3	0.50	0.011	0.094	0.094	0.380	4.500
4	0.50	0.023	0.125	0.125	3.139	17.500
5	0.50	0.023	0.156	0.156	17.031	40.000
6	0.50	0.047	0.188	0.188	5.839	25.500
7	0.50	0.060	0.219	0.219	0.779	6.500
8	1.00	0.064	0.250	0.250	3.806	19.000
9	1.00	0.067	0.281	0.281	13.799	40.000
10	1.00	0.085	0.313	0.313	0.323	3.000
11	1.50	0.094	0.344	0.344	1.044	8.000
12	3.00	0.285	0.375	0.375	0.104	1.500
13	4.50	0.457	0.406	0.406	2.540	17.500
14	6.50	0.667	0.438	0.438	14.144	40.000
15	8.00	1.202	0.469	0.469	0.056	0.500
16	8.50	1.366	0.500	0.500	0.074	1.000
17	11.00	1.639	0.531	0.531	0.033	0.500
18	15.00	1.685	0.563	0.563	11.273	39.500
19	15.50	1.718	0.594	0.594	**23.524**	**57.500**
20	16.00	1.925	0.625	0.625	0.472	4.500
21	17.50	2.387	0.656	0.656	0.403	4.500
22	19.00	4.121	0.688	0.688	0.047	0.500
23	25.00	5.042	0.719	0.719	0.038	0.500
24	25.50	5.157	0.750	0.750	0.137	1.500
25	26.50	8.656	0.781	0.781	0.013	0.500
26	38.50	10.588	0.813	0.813		
27	39.50	10.709	0.844	0.844		
28	40.00	15.233	0.875	0.875		
29	55.00	19.541	0.906	0.906		
30	57.50	21.872	0.938	0.938		
31	115.00	76.989	0.969	0.969		

[a] Observation data in the second column of Table 4.3 was sorted in increasing order.
[b] GCMcur data in the second column of Table 4.3 was sorted in increasing order.

observation that matches F_m for the bias-corrected value (i.e., 57.50 mm). The bias-corrected GCM future values are presented in the last column of Table 4.4. See also Figure 4.1 for illustration of this example.

4.4.3 Quantile Delta Mapping

It is critical to preserve model-projected relative changes in the bias correction. Cannon et al. (2015) suggested quantile delta mapping (QDM) to preserve the relative changes of model projections. The bias correction with QDM is performed with the dataset in Tables 4.3 and 4.4 and is presented in Table 4.5.

For the model projected future series ($x_{m,fut}$), the ECDF of this series around the projected period is estimated as

$$\tau_{m,fut} = F_{m,fut}\left(x_{m,fut}\right) \tag{4.9}$$

The relative change in quantiles between the base period and the future projection period is given by

$$\Delta_{m,fut} = \frac{F_{m,fut}^{-1}\left(x_{m,fut}\right)}{F_{m,base}^{-1}\left(\tau_{m,fut}\right)} = \frac{x_{m,fut}}{F_{m,base}^{-1}\left(\tau_{m,fut}\right)} \tag{4.10}$$

TABLE 4.5
Example of QDM with Empirical Distribution with the Dataset in Table 4.3

Ord	GCMfut, $x_{m,fut}$	GCMfut ord.[a]	$F_{m,fut}$	$\tau_{m,fut}$[b]	$F_{m,base}^{-1}\left(t_{m,fut}\right)$	$\Delta_{m,fut}$	$\hat{y}_{o,fut}$	$\hat{y}_{fut}$
1	3.739	0.013	0.038	0.692	4.121	0.907	19.000	17.240
2	1.741	0.033	0.077	0.577	1.685	1.033	15.000	15.496
3	0.380	0.038	0.115	0.385	0.285	1.333	3.000	3.998
4	3.139	0.047	0.154	0.654	2.387	1.315	17.500	23.014
5	17.031	0.056	0.192	0.923	21.872	0.779	57.500	44.773
6	5.839	0.074	0.231	0.769	8.656	0.675	26.500	17.876
7	0.779	0.104	0.269	0.500	1.366	0.571	8.500	4.850
8	3.806	0.137	0.308	0.731	5.042	0.755	25.000	18.872
9	13.799	0.323	0.346	0.846	10.709	1.289	39.500	50.897
10	0.323	0.380	0.385	0.346	0.094	3.437	1.500	5.155
11	1.044	0.403	0.423	0.538	1.639	0.637	11.000	7.009
12	0.104	0.472	0.462	0.269	0.067	1.548	1.000	1.548
13	2.540	0.779	0.500	0.615	1.925	1.320	16.000	21.112
14	14.144	1.044	0.538	0.885	15.233	0.929	40.000	37.140
15	0.056	1.741	0.577	0.192	0.047	1.188	0.500	0.594
16	0.074	2.540	0.615	0.231	0.060	1.236	0.500	0.618
17	0.033	3.139	0.654	0.077	0.010	3.261	0.200	0.652
18	11.273	3.739	0.692	0.808	10.588	1.065	38.500	40.991
19	23.524	3.806	0.731	0.923	21.872	1.076	57.500	61.842
20	0.472	5.839	0.769	0.462	1.202	0.393	8.000	3.144
21	0.403	11.273	0.808	0.423	0.667	0.605	6.500	3.931
22	0.047	13.799	0.846	0.154	0.023	2.058	0.500	1.029
23	0.038	14.144	0.885	0.115	0.023	1.671	0.500	0.836
24	0.137	17.031	0.923	0.308	0.085	1.607	1.000	1.607
25	0.013	23.524	0.962	0.038	0.009	1.489	0.200	0.298

[a] The GCM model values at the second column are sorted in increasing order.

[b] $F_{m,fut}^{*}$ is just $F_{m,fut}$the third column but the corresponding ECDF of the GCM data in the second column.

The modeled $\tau_{m,fut}$ quantile can be bias-corrected by applying the inverse ECDF of the observations as

$$\hat{y}_{o,fut} = F_o^{-1}\left(\tau_{m,fut}\right) \tag{4.11}$$

Finally, the target future model value can be estimated as

$$\hat{y}_{m,fut} = \Delta_{m,fut}\hat{y}_{o,fut} \tag{4.12}$$

This is explicitly expressed as

$$\hat{y}_{m,fut} = \frac{F_o^{-1}\left[F_{m,fut}\left(x_{m,fut}\right)\right]}{F_{m,base}^{-1}\left[F_{m,fut}\left(x_{m,fut}\right)\right]}x_{m,fut} = \delta \cdot x_{m,fut} \tag{4.13}$$

where $\delta = \frac{F_o^{-1}\left[F_{m,fut}\left(x_{m,fut}\right)\right]}{F_{m,base}^{-1}\left[F_{m,fut}\left(x_{m,fut}\right)\right]}$. This indicates that the QDM adjusts the original model output ($x_{m,fut}$) with the ratio of the inverse of observational CDF $\left(F_o^{-1}[\cdot]\right)$ and the base-period model CDF $\left(F_{m,base}^{-1}[\cdot]\right)$ for the projection-period model CDF of the original model output ($F_{m,fut}(x_{m,fut})$). The ECDF can be easily adopted for any of the CDFs.

Example 4.5

Correct the bias of the GCM future precipitation data in the second column of Table 4.5 with QDM. Note that the ECDFs of observation and the model output for the base period are in the fourth and fifth columns.

Employing the ECDF for all the CDFs (F_o, $F_{m,base}$, $F_{m,fut}$), the procedure is as follows:

1. Sort the GCM future data as shown in the third column of Table 4.5
2. Estimate the corresponding ECDF ($F_{m,fut}$) as in Eq. (4.8), i.e., $i/(25 + 1)$, where $i = 1,\ldots, 25$. Note that the ECDF values in the fourth column, $\tau_{m,fut}$, present the ECDF corresponding to the GCM future values in the second column. For example, the ECDF ($\tau_{m,fut}$) of the first value of the future GCM in the second column ($x_{m,fut} = 3.739$ mm) can be calculated from the GCM data ordered in the second column and its corresponding ECDF ($F_{m,fut}$ in the third column), i.e., $\tau_{m,fut} = 0.692$.
3. Estimate $F_{m,base}^{-1}\left(\tau_{m,fut}\right)$ for the CDF $\tau_{m,fut}$ in the fifth column of Table 4.5 with the ECDF of $F_{m,base}$ in the fifth column of Table 4.4. For example, the inverse ECDF value $F_{m,base}^{-1}\left(\tau_{m,fut}\right)$ for $\tau_{m,fut} = 0.692$ is estimated by finding the $F_{m,base}$ closest to the $\tau_{m,fut} = 0.692$ in the fifth column of Table 4.4, which is $F_{m,base} = 0.688$ (22nd). The corresponding ordered GCM base value ($x_{m,base}$) in the third column of Table 4.4 is 4.121 mm.

4. Calculate $\Delta_{m,fut}$ as $x_{m,fut}/F_{m,base}^{-1}(\tau_{m,fut})$. For example, $\Delta_{m,fut} = 3.739/4.121 = 0.907$.
5. Estimate $\hat{y}_{o,fut}$ for $\tau_{m,fut}$ with the observed ECDF (F_o) in the fourth column of Table 4.4. For example ($\tau_{m,fut}$ = 0.692), the observed ECDF closest to $\tau_{m,fut}$ = 0.692 is 0.688 and its corresponding observation is 19.00 mm, i.e., $\hat{y}_{o,fut} = F_o^{-1}(0.692) = 19.00$ mm.
6. Calculate the final target project GCM as $\hat{y}_{m,fut} = \hat{y}_{o,fut}\Delta_{m,fut}$. For example, $\hat{y}_{m,fut} = \hat{y}_{o,fut}\Delta_{m,fut} = 19.00 \times 0.907 = 17.240$. See the last column of Table 4.5 for all the downscaled data with QDM.

Instead of multiplicative delta for QDM in Eq. (4.10), the additive delta can be applied as

$$\Delta_{m,fut} = x_{m,fut} - F_{m,base}^{-1}(\tau_{m,fut}) \tag{4.14}$$

The target future GCM can be estimated as

$$\hat{y}_{m,fut} = \hat{y}_{o,fut} + \Delta_{m,fut} = \hat{y}_{o,fut} + x_{m,fut} - F_{m,base}^{-1}(\tau_{m,fut}) \tag{4.15}$$

For example,

$$\Delta_{m,fut} = 3.739 - 4.121 = -0.382$$

and

$$\hat{y}_{m,fut} = \hat{y}_{o,fut} + \Delta_{m,fut} = 19.00 - 0.382 = 18.618$$

4.5 SUMMARY AND COMPARISON

Different approaches of correcting biases of systematic errors in GCM outputs are discussed in this section as well as discussion of the necessity of bias correction and the occurrence adjustment for precipitation. The delta method employs the ratios of the key statistics as mean and standard deviation and adjust target GCM outputs employing these ratios in Eq. (4.2). QM uses the CDF of GCM outputs for a base period and its inverse function of observations, with different marginal distributions such as exponential, gamma, and empirical as in Eq. (4.7). A recent bias correction method, QDM, employing the ratios of the CDFs of future and base periods is presented, as in Eq. (4.13). These bias correction methods are applied to the selected dataset, as shown in Table 4.2.

In Figure 4.2, the time series of the original future GCM output data (black thick line) and bias-corrected series are presented. It is noticed that the original future GCM is relatively smaller than the BC series. The BC series with the delta method tend to be lower than the other methods, while the series with the QDM seem relatively higher. However, this is just a simple case that cannot be easily generalized with simple case studies.

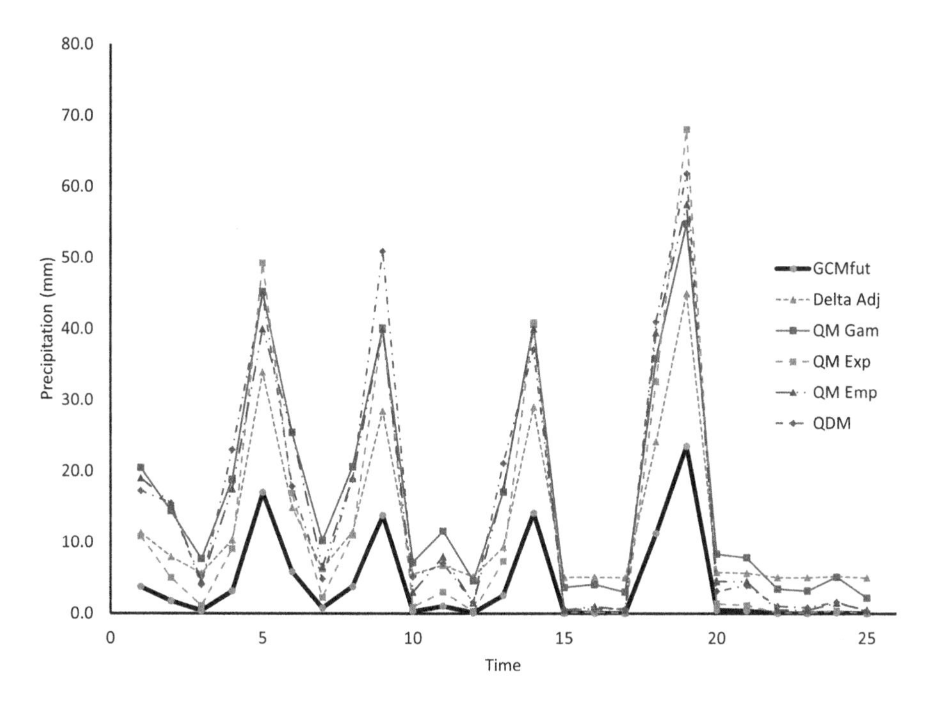

FIGURE 4.2 **(See color insert.)** Time series of the original future GCM output data (black thick line with gray circle marker) and bias-corrected series: (1) Delta method (dotted light blue line with triangle marker), (2) QM with gamma distribution (solid blue line with square marker), (3) QM with exponential distribution (dashed green line with square marker), (4) QM with empirical CDF (dash-dotted blue line with triangle marker), (4) QDM with empirical CDF (dash-dotted red line with diamond marker).

A selection of an appropriate BC method has been discussed in the literature (Maraun, 2016) and is rather subjective according to its complexity and objective. For example, when the BC precipitation is used for estimating designs of hydraulic structures, extreme precipitation values must be reliable rather than small values. It is totally different when BC series is used in drought analysis. Therefore, which method is used is highly subjective and locally different, even though each BC approach has its own characteristics. Statistical tests must be applied to the BC series according to the objective of interests.

5 Regression Downscalings

5.1 LINEAR REGRESSION BASED DOWNSCALING

One of the most common ways to downscale GCM future projections to a point of interest (or station) is to employ regression methods. Three different datasets must be acquired for regression-based downscaling: (1) observation data for variables of interest (predictand), (2) reanalysis data of predictor variables, and (3) GCM output for predictor variables. Here, reanalysis data is the produced record of global analyses of atmospheric fields by using a frozen state-of-the-art analysis/forecast system and performing data assimilation using past data. Those produced atmospheric variables (i.e., reanalysis data) can be reliably compared with GCM outputs.

A regression-based model is commonly employed to downscale GCM outputs to a site where a weather station is located nearby. As shown in Figure 5.1, the procedure is as follows:

Fit a regression model to the observed variable of interest (Y) and predictor variables (X) of reanalysis dataset. Note that one or more predictor variables can be used.

Apply the predictor variables of GCM outputs to the fitted regression model from step (1) instead of those of the reanalysis dataset.

Obtain the future projection of the variable of interest.

5.1.1 Simple Linear Regression

A simple linear model for the deterministic case is described as $y = \beta_0 + \beta_1 x$, which is the actual observed value of y as a linear function of x with parameters β_0 and β_1. This linear model can be generalized to a probabilistic model for the random variable Y as

$$Y = \beta_0 + \beta_1 X + \varepsilon \tag{5.1}$$

Note that Y is capitalized, because the output of the model includes a random noise (ε) assumed to be normally distributed with $E(\varepsilon) = 0$ and $Var(\varepsilon) = \sigma_\varepsilon^2$, and the output is not a fixed value any more but is a random variable. Also, X is called a predictor, while Y is a predictand (or an explanatory variable).

For a sample of observed data pairs of size n, i.e., (x_i, y_i) for $i = 1, \ldots, n$, the sum of square of errors (SSE) is defined as

$$SSE = \sum_{i=1}^{n} \left[y_i - (\beta_0 + \beta_1 x_i) \right]^2 = \sum_{i=1}^{n} \varepsilon_i^2 \tag{5.2}$$

The parameters can be estimated by finding the parameter set that provides the minimum quantity of SSE, called least-square estimate, which is analogous to taking the derivative of SSE with respect to the parameters and equating the derivatives to zero as

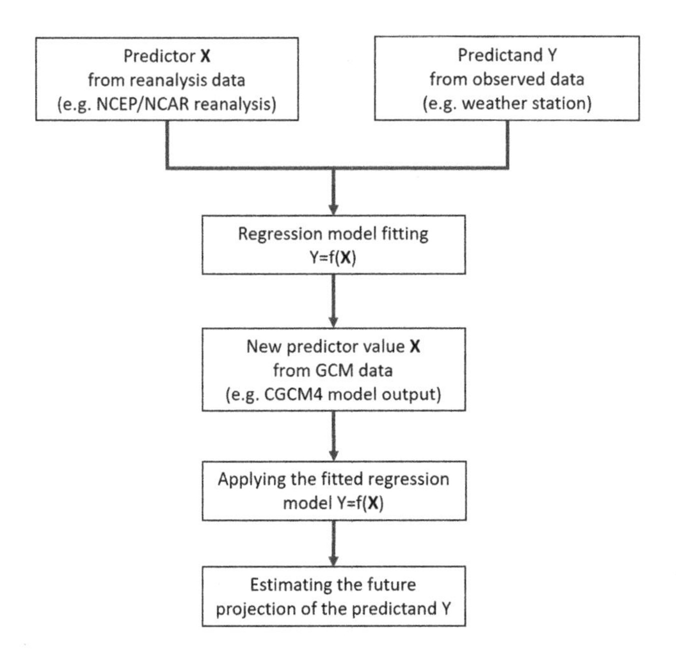

FIGURE 5.1 Procedure of the regression-based downscaling. Note that in fitting a regression model including parameter estimation.

$$\frac{\partial SSE}{\partial \beta_0} = -2\sum_{i=1}^{n}\left[y_i - (\beta_0 + \beta_1 x_i)\right] = 0 \tag{5.3}$$

$$\frac{\partial SSE}{\partial \beta_1} = -2\sum_{i=1}^{n}\left[y_i - (\beta_0 + \beta_1 x_i)\right](x_i) = 0 \tag{5.4}$$

Then, from these two equations, the least-square estimate of β_0 and β_1, denoted as $\hat{\beta}_0, \hat{\beta}_1$, can be derived as

$$\hat{\beta}_1 = \frac{\sum_{i=1}^{n}(x_i - \bar{x})(y_i - \bar{y})}{\sum_{i=1}^{n}(x_i - \bar{x})^2} \tag{5.5}$$

$$\hat{\beta}_0 = \bar{y} - \hat{\beta}_1 \bar{x} \tag{5.6}$$

The variance estimate of $\hat{\beta}_1$ is

$$s_{\hat{\beta}_1}^2 = \frac{s_\varepsilon^2}{\sum_{i=1}^{n}(x_i - \bar{x})^2} \tag{5.7}$$

and

$$s_{\varepsilon}^{2} = \frac{\sum_{i=1}^{n}\left[y_i - \left(\hat{\beta}_0 + \hat{\beta}_1 x_i\right)\right]^2}{n-2} = \frac{\sum_{i=1}^{n}\left[y_i - \hat{y}_i\right]^2}{n-2} = \frac{\sum_{i=1}^{n}\varepsilon_i^2}{n-2} \tag{5.8}$$

where $\hat{y}_i$ is the estimate of y_i. Note that the divisor n–2 which is the number of degrees of freedom associated with the estimate and two parameters is used to obtain s_{ε}^{2}, resulting in a loss of 2 degrees of freedom (Devore, 1995).

5.1.1.1 Significance Test

It has been shown that the following standardized variable T has a t-distribution with n–2 degrees of freedom:

$$T = \frac{\hat{\beta}_1 - \beta_1}{s_{\hat{\beta}_1}} \tag{5.9}$$

This variable is used for a hypothesis test that $\beta_1 = 0$. Comparing the t-distribution critical value for the significance level α ($t_{\alpha/2,n-2}$) with variable T,

Null hypothesis H_0: $\beta_1 = 0$

Test statistic value: $T = \frac{\hat{\beta}_1 - 0}{s_{\hat{\beta}_1}} = \frac{\hat{\beta}_1}{s_{\hat{\beta}_1}}$

Alternative hypothesis: $H_a : \beta_1 \neq 0$

Rejection region for the significance level: $T \geq t_{\alpha/2,n-2}$

When the estimated test statistic value T is larger than the t-distribution critical value of ($t_{\alpha/2,n-2}$), the null hypothesis that $H_0 : \beta_1 = 0$ is rejected and the alternative hypothesis that $H_a : \beta_1 \neq 0$ is supported. In other words, the predictor variable X has a strong relationship with Y.

In addition, the p-value that determines the probability that the value of T estimated with the assumption that the null hypothesis H_0 might occur can be calculated from the distribution as $p = 2[1-Pr(T \leq t)]$. The applied distribution is the same t-distribution with n–2 degrees of freedom.

Example 5.1

Downscale the monthly average of daily maximum temperature for January at Cedars, Quebec, using a simple linear regression with the predictor of near-surface specific humidity of National Centers for Environmental Prediction/National Center for Atmospheric Research (NCEP/NCAR). All the required dataset is shown in Table 5.1.

Solution:

Three datasets are employed as

a. Monthly average (January) of the daily maximum temperature (°C) at Cedars (latitude: 45.3°, longitude:–74.05°), Canada (http://climate.weather.gc.ca/historical_data/search_historic_data_e.html)

TABLE 5.1
Example of Simple Linear Regression with Monthly Average of Near-Surface Specific Humidity (*x*, Predictor; Unit: g/kg) and Daily Maximum Temperature (*y*, Predictand; Unit: °C) for January

	x_i	y_i(°C)	$(x_i - \bar{x})^2$	$(y_i - \bar{y})(x_i - \bar{x})$	$\hat{y}_i$	$(y_i - \hat{y}_i)^2$
1961[a]	1.463	−9.106	0.267	1.746	−8.768	0.115
1962	1.765	−6.319	0.046	0.127	−6.994	0.455
1963	1.611	−5.490	0.136	−0.089	−7.898	5.799
1964	2.368	−1.819	0.151	1.517	−3.451	2.662
1965	1.503	−5.726	0.228	−0.002	−8.533	7.880
1967	2.731	−1.165	0.564	3.429	−1.318	0.024
1968	1.634	−8.761	0.120	1.049	−7.763	0.996
1969	2.156	−4.394	0.031	0.235	−4.696	0.092
1970	1.328	−10.245	0.425	2.944	−9.561	0.468
1971	1.600	−8.545	0.144	1.070	−7.963	0.339
1972	2.283	−2.245	0.092	1.056	−3.950	2.908
1973	2.650	−3.635	0.449	1.403	−1.794	3.391
1974	2.231	−4.681	0.063	0.263	−4.256	0.180
1975	2.476	−3.000	0.246	1.354	−2.816	0.034
1976	1.653	−9.242	0.107	1.148	−7.652	2.529
1977	1.456	−9.303	0.275	1.872	−8.809	0.244
1978	1.779	−7.694	0.040	0.395	−6.911	0.612
1979	2.232	−5.342	0.063	0.098	−4.250	1.192
1980	1.952	−4.097	0.001	−0.046	−5.895	3.234
1981	1.369	−10.435	0.373	2.875	−9.320	1.244
1984	1.769	−7.774	0.045	0.431	−6.970	0.646
1985	1.543	−8.677	0.191	1.288	−8.298	0.144
1986	2.316	−4.565	0.113	0.392	−3.756	0.653
1987	2.178	−4.645	0.039	0.215	−4.567	0.006
1989	2.386	−2.129	0.165	1.462	−3.345	1.479
1990	3.049	0.048	1.143	6.177	0.550	0.252
Sum	51.481	−148.987	5.516	32.410	−148.987	37.576

[a] Some years with missing data are not used, e.g., 1966.

b. Monthly average (January) of near-surface specific humidify (g/kg) of NCEP/NCAR reanalysis data (www.esrl.noaa.gov/psd/data/gridded/data.ncep.reanalysis.pressure.html)
c. Monthly average (January) of near-surface specific humidify (g/kg) of Canadian Global Climate Model (CGCM) output (www.cccma.ec.gc.ca/data/cgcm4/CanESM2/index.shtml)

In Table 5.1, the target predictand variable (*Y*) is the monthly maximum temperature (°C) for January in the third column, and its predictor is the monthly average of near-surface specific humidity (g/kg) in the second column.

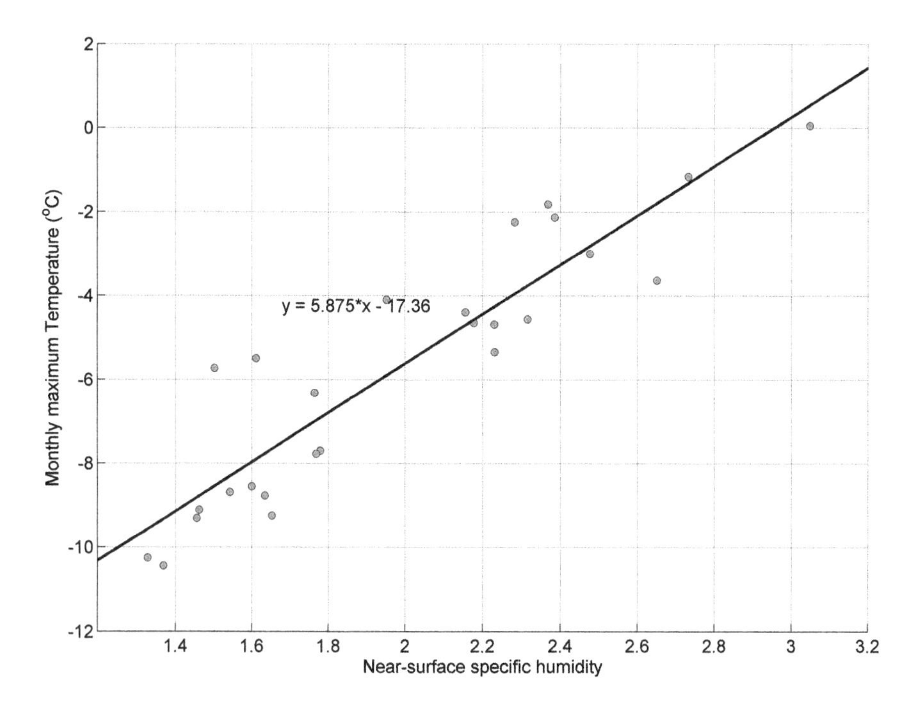

FIGURE 5.2 Downscaling of monthly maximum surface temperature (°C) with the simple linear regression employing near-surface specific humidity as a predictor. Note that only the January data were employed to avoid the seasonality of the dataset.

The scatterplot of these two variables is shown in Figure 5.2. The scatterplot presents that daily temperature has a positive relationship with the near-surface specific humidity, because the fitted line shows that the temperature variable is increased along with increasing specific humidity.

The linear regression model parameters (i.e., β_0, β_1) can be estimated with Eqs. (5.6) and (5.5), respectively.

Here, $\bar{x} = 51.481/26 = 1.980$ and $\bar{y} = -148.987/26 = -5.730$

$$\hat{\beta}_1 = \frac{\sum_{i=1}^{n}(x_i - \bar{x})(y_i - \bar{y})}{\sum_{i=1}^{n}(x_i - \bar{x})^2} = \frac{32.410}{5.516} = 5.875$$

$$\hat{\beta}_0 = \bar{y} - \hat{\beta}_1\bar{x} = -5.730 - (5.875 \times 1.98) = -17.363$$

The significance test can be performed for $\hat{\beta}_1$ for the null hypothesis: $H_0 : \beta_1 = 0$.

$$s_\varepsilon^2 = \frac{\sum_{i=1}^{n}[y_i - \hat{y}_i]^2}{n-2} = \frac{\sum_{i=1}^{n}\varepsilon_i^2}{n-2} = \frac{37.576}{24} = 1.566$$

$$s_{\hat{\beta}_1}^2 = \frac{s_\varepsilon^2}{\sum_{i=1}^{n}(x_i - \bar{x})^2} = \frac{1.566}{5.516} = 0.284$$

$$T = \frac{\hat{\beta}_1 - \beta_1}{s_{\hat{\beta}_1}} = \frac{5.875 - 0}{\sqrt{0.284}} = 11.028$$

$$t_{\alpha/2,n-2} = t_{0.025,24} = 2.064$$

Note that typically 95% confidence interval is employed and the significance level for both sides (i.e., $\alpha/2$) is employed for the consideration of $\beta_1 > 0$ and $\beta_1 < 0$, and the confidence interval is

$$P\left(-t_{\alpha/2,n-2} < \frac{\hat{\beta}_1 - \beta_1}{s_{\hat{\beta}_1}} < t_{\alpha/2,n-2}\right) = 1 - \alpha = 0.95$$

Since $T \geq t_{\alpha/2,n-2}$ as $T = \frac{\hat{\beta}_1 - \beta_1}{s_{\hat{\beta}_1}} = 11.028$ and $11.028 \geq 2.064$, the null hypothesis is rejected for the 95% confidence level, implying that the linear relation between two variables is statistically significant. This test is critical in the linear regression downscaling in that there is a significant relation with the predictor variable to use the current linear regression model. The p-value can be estimated as $2[1-Pr(T < t)] = 7.03e^{-11}$.

The future projection of monthly maximum temperature for the Cedars site is obtained by applying the near-surface specific humidity of the CGCM output. Note that the data employed in this fitting is the NCEP/NCAR reanalysis data. One may refer to Chapter 3 for the difference between reanalysis data and GCM output. The partial dataset of the employed specific humidity (X) of the CGCM output as well as its estimated value for the base and future projections of the monthly maximum temperature are presented in Table 5.2.

In Figure 5.3, the downscaled temperatures for the base period (dotted line) of 1850–2005 and the future period (dotted line with cross markers) of 2006–2100 are illustrated by employing the near-surface specific humidity of CGCM for the January average as well as the observed (solid line with circles) monthly maximum temperature. A significant increase is observed in the downscaled future projection. The record period of the observed data is 1960–1990 with some missing years. It is common that the observed data is often short and might be problematic, and its reliability is still one of the concerns.

5.1.2 Multiple Linear Regression

The multiple regression for multiple variables of $x_i^1, x_i^2, \ldots, x_i^S$ is described as

$$Y_i = \beta_0 + \beta_1 x_i^1 + \beta_2 x_i^2 + \cdots + \beta_s x_i^s + \varepsilon_i \tag{5.10}$$

where S is the number of predictors of interest. In matrix form, the linear regression model is expressed by

TABLE 5.2
Partial Dataset of the CGCM Predictor and Their Predicted Values with the Parameter Set in Figure 5.2

Base Period			Future Period		
Year	**X**	$\hat{Y}$	**Year**	**X**	$\hat{Y}$
1991	1.861	−6.427	2086	3.278	1.897
1992	1.591	−8.019	2087	2.058	−5.272
1993	2.520	−2.558	2088	2.746	−1.230
1994	2.373	−3.421	2089	2.614	−2.006
1995	1.736	−7.165	2090	3.766	4.760
1996	1.999	−5.622	2091	2.291	−3.901
1997	1.473	−8.711	2092	3.360	2.375
1998	2.581	−2.199	2093	3.899	5.545
1999	1.702	−7.367	2094	3.158	1.189
2000	1.668	−7.564	2095	3.729	4.547
2001	1.812	−6.716	2096	2.746	−1.229
2002	1.557	−8.214	2097	2.362	−3.487
2003	2.321	−3.730	2098	3.169	1.255
2004	1.819	−6.677	2099	2.470	−2.853
2005	1.567	−8.157	2100	2.294	−3.884

$$Y_i = \vec{\mathbf{x}}_i^T \boldsymbol{\beta} + \varepsilon_i \tag{5.11}$$

where $\vec{\mathbf{x}}_i = [1, x_i^1, x_i^2, \ldots, x_i^S]^T$ and $\boldsymbol{\beta} = [\beta_0, \beta_1, \ldots, \beta_s]^T$.

The sum of squares of residuals and its least-square estimate of $\boldsymbol{\beta}$ are given by

$$SSE = \sum_{t=1}^{n} (\varepsilon_t)^2 = \mathbf{e}'\mathbf{e} = (\mathbf{y} - \vec{\mathbf{x}}\boldsymbol{\beta})^T(\mathbf{y} - \vec{\mathbf{x}}\boldsymbol{\beta})$$

$$= \mathbf{y}^T\mathbf{y} - \mathbf{y}^T\vec{\mathbf{x}}\boldsymbol{\beta} - \boldsymbol{\beta}^T\vec{\mathbf{x}}^T\mathbf{y} + \boldsymbol{\beta}^T\vec{\mathbf{x}}^T\vec{\mathbf{x}}\boldsymbol{\beta}$$

where $\mathbf{e} = [\varepsilon_1, \ldots, \varepsilon_n]^T$, $\vec{\mathbf{x}} = [1, \ldots, 1; x_1^1, \ldots, x_n^1; x_1^2, \ldots, x_n^2; \ldots; x_1^S, \ldots, x_n^S]^T$ ($n \times S + 1$ matrix), and $\mathbf{y} = [y_1, \ldots, y_n]^T$ with n number of data. The derivative to have its minimum with $dSS/d\boldsymbol{\beta} = 0$ can be obtained as Eq. (5.12):

$$\frac{dSSE}{d\boldsymbol{\beta}} = -2\vec{\mathbf{x}}^T\mathbf{y} + 2\vec{\mathbf{x}}^T\vec{\mathbf{x}}\boldsymbol{\beta} = 0$$

The least-square estimator is obtained as

$$\hat{\boldsymbol{\beta}} = (\vec{\mathbf{x}}^T\vec{\mathbf{x}})^{-1}\vec{\mathbf{x}}^T\mathbf{y} \tag{5.12}$$

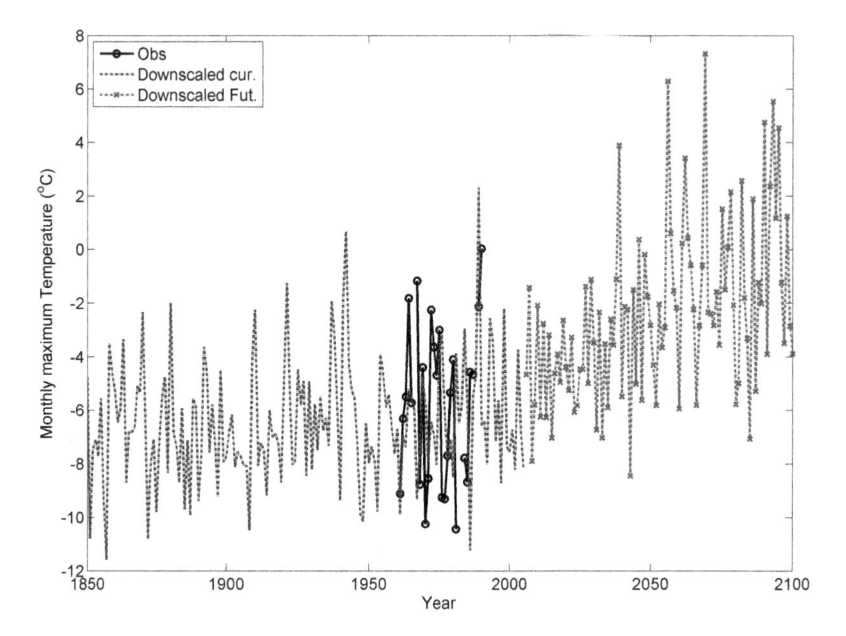

FIGURE 5.3 Downscaled temperature for the base period (dotted line) of 1850–2005 and the future period (dotted line with cross markers) of 2006–2100 employing the near-surface specific humidity of CGCM for the January average as well as the observed (solid line with circles) monthly maximum temperature. See Tables 5.1 and 5.2 for the detailed values.

Source: **File name of the base period: Models/CanESM2/historical/mon/atmos/huss/r1i1p1/huss_Amon_CanESM2_historical_r1i1p1_185001–200512.nc; www.cccma.ec.gc.ca/data/cgcm4/CanESM2/historical/mon/atmos/huss/index.shtml; File name of the future scenario: Models/CanESM2/rcp45/mon/atmos/huss/r1i1p1/huss_Amon_CanESM2_rcp45_r1i1p1_200601–210012.nc; www.cccma.ec.gc.ca/data/cgcm4/CanESM2/rcp45/mon/atmos/huss/index.shtml**

Also, the error variance $Var(\varepsilon) = \sigma_{\varepsilon}^2$ is estimated with

$$\hat{\sigma}_{\varepsilon}^2 = \frac{SSE}{n-(k+1)} \tag{5.13}$$

Example 5.2

Downscale the same dataset as in Example 5.1 but with multiple regression, including the additional predictor variables summarized in Table 5.3 as sea-level pressure (SLP), 850 hPa geopotential height (GPH), specific humidity, and surface air temperature.

Solution:

The employed data values are shown in Table 5.4. Their relations with the predictand variable of maximum temperature are presented in Figure 5.4. A strong linear correlation with the predictand variable can be observed for the predictor variables of specific humidity and surface air temperature.

TABLE 5.3
Information of the Dataset Employed for Multiple Linear Regression

	Variable Name		Unit	
			NCEP/NCAR[a]	**CGCM RCP4.5**[d]
Employed Predictor Variables	SLP	hPa	Pa	
	850 hPa GPH	m	m	
	Specific humidity	g/kg	g/g	
	Near-surface air temperature	°C	°K	
	Dataset	Obs	NCEP/NCAR[c]	CGCM RCP4.5[d]
Location	Latitude (°)	45.3	45 (19[a])	46.045 (49)
	Longitude (°)	285.95	285 (115)	286.875 (103)
Record	Monthly mean of January	1961–1990[b]	1961–1990[b]	1850–2005 (base) 2006–2100 (future)

Note that all the variables are monthly average value and the January data is only employed to avoid the *seasonality.*

[a] The number in the parenthesis presents the grid number of the downloaded data matrix.
[b] The data with missing values were not included, such as 1966, 1982, 1983, and 1988.
[c] NCEP/NCAR data were downloaded from www.esrl.noaa.gov/psd/data/gridded/data.ncep.reanalysis.html.
[d] CGCM data were downloaded from www.cccma.ec.gc.ca/data/cgcm4/CanESM2/index.shtml.

Note that the NCEP/NCAR dataset and CGCM output have different units so that it requires a unit conversion. Also, the data scales of the variables employed are quite different from each other (e.g., SLP: 10^3 and 10^5 hPa and specific humidity: 1 and 10^{-3} g/kg). A variable with a larger scale is more influential to a predictor in multiple linear regression than with smaller scale. To avoid this scale issue, the preprocessing of standardization can be applied with the following procedure:

a. Each variable including the predictor is standardized as

$$x_i = \frac{(x_i^{org} - \bar{x})}{s_x} \tag{5.14}$$

where x_i^{org} is the variable that is in the original domain (i.e., before standardization)

For example, the SLP of January 1961 is 1017.07 hPa and its mean and standard deviation are 1015.62 and 2.87 hPa, respectively. The standardized value is x_i = (1017.07–1015.62)/2.87 = 0.504. Also, the target predictor Y for January 1961 value is −8.575°C and its mean and standard deviation are −5.73 and 2.87°C, respectively. The standardized value of the predictor (monthly maximum temperature at Cedars, Quebec) on January 1961 is y_i = (−8.575–(−5.73))/3.02 = −1.118. All the standardized data are shown in Table 5.5.

TABLE 5.4
Example of Dataset for Fitting the Multiple Linear Regression Model

	Employed Predictors (NCEP/NCAR Reanalysis)						
	SLP (hPa)	850 hPa GPH (m)	Specific Humidity (g/kg)	Near-Surf. Air Temp. (°C)	Monthly Max Temp. (Y; °C)	$\hat{y}_i$	$\varepsilon^2 = (y_i - \hat{y}_i)^2$
1961	1017.07	1372.39	1.463	−13.538	−9.106	−8.575	0.282
1962	1017.20	1377.81	1.765	−11.341	−6.319	−6.417	0.010
1963	1017.51	1380.97	1.611	−11.857	−5.490	−6.732	1.542
1964	1014.12	1376.26	2.368	−7.830	−1.819	−3.030	1.465
1965	1016.74	1371.87	1.503	−12.783	−5.726	−7.724	3.992
1967	1014.37	1383.23	2.731	−6.112	−1.165	−1.534	0.136
1968	1023.09	1428.06	1.634	−13.430	−8.761	−8.377	0.148
1969	1017.55	1401.16	2.156	−9.041	−4.394	−3.999	0.156
1970	1015.21	1351.77	1.328	−14.809	−10.245	−10.024	0.049
1971	1012.21	1333.87	1.600	−12.982	−8.545	−8.347	0.039
1972	1015.74	1379.35	2.283	−8.757	−2.245	−4.090	3.403
1973	1015.02	1383.45	2.650	−7.632	−3.635	−3.261	0.140
1974	1018.11	1399.45	2.231	−9.010	−4.681	−4.183	0.247
1975	1015.73	1387.55	2.476	−7.137	−3.000	−2.360	0.410
1976	1016.24	1368.55	1.653	−13.072	−9.242	−8.344	0.807
1977	1009.18	1303.00	1.456	−14.210	−9.303	−9.793	0.240
1978	1014.73	1358.97	1.779	−11.866	−7.694	−7.151	0.295
1979	1014.08	1361.94	2.232	−9.229	−5.342	−4.702	0.410
1980	1017.16	1388.48	1.952	−10.523	−4.097	−5.584	2.211
1981	1015.19	1348.87	1.369	−15.426	−10.435	−10.882	0.199
1984	1020.65	1406.97	1.769	−11.261	−7.774	−6.162	2.601
1985	1010.67	1325.03	1.543	−12.247	−8.677	−7.371	1.705
1986	1014.77	1371.10	2.316	−9.260	−4.565	−4.796	0.053
1987	1013.62	1369.23	2.178	−8.350	−4.645	−3.383	1.592
1989	1017.53	1404.71	2.386	−7.485	−2.129	−2.492	0.132
1990	1012.68	1380.45	3.049	−4.177	0.048	0.324	0.076
Mean	1015.62	1373.63	1.98	−10.51	−5.73	−5.73	0.86
Std	2.87	26.57	0.47	2.90	3.02	2.87	1.12

Note that the employed dataset of the target predictand variable is the monthly maximum (max) *temperature of January at Cedars, Quebec. The dataset of the employed predictors was extracted from NCEP/NCAR reanalysis data, while the observed monthly maximum temperature was from Environment Canada.*

b. Multiple linear regression model is fitted to the standardized data (x_i) instead of x_i^{org}. The parameter vector is estimated with Eq. (5.12) for this example as

$$\hat{\boldsymbol{\beta}} = [0, \ -0.096, \ 0.156, \ -0.278, \ 1.144]^T$$

Note that $\hat{\beta}_0 = 0$, since all the mean estimates are zero with the standardized value. See Eq. (5.6) for a simple case as $\hat{\beta}_0 = \bar{y} - \hat{\beta}_1 \bar{x} = 0 - \hat{\beta}_1 0 = 0$. The

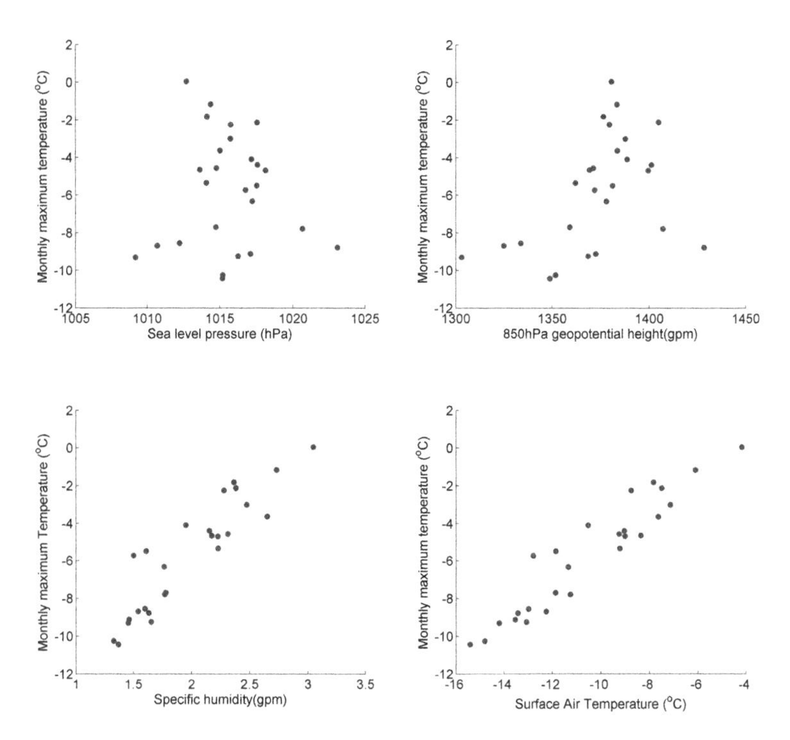

FIGURE 5.4 Scatterplot between the target monthly maximum temperature (°C) at Cedars, Quebec, and the four atmospheric predictors obtained from NCEP/NCAR reanalysis data with multiple linear regression. Note that only the January data were employed to avoid the seasonality of the dataset.

predictor estimate is then made by applying Eq. (5.11) setting the noise term as zero (i.e., $E(\varepsilon) = 0$), then

$$\hat{y}_i = \vec{\mathbf{x}}_i^T \hat{\boldsymbol{\beta}} \tag{5.15}$$

The value for January 1961 is estimated as

$$\begin{bmatrix} 1 \\ 0.504 \\ -0.047 \\ -1.101 \\ -1.042 \end{bmatrix}^T \times \begin{bmatrix} 0 \\ -0.096 \\ 0.156 \\ -0.278 \\ 1.144 \end{bmatrix} = -0.942.$$

Here, the vector multiplication is applied.

c. The estimate from the multiple linear regression with the standardized data is back-standardized as

$$\hat{y}_i^{org} = \hat{y}_i \times s_y + \bar{y}$$

TABLE 5.5
Standardized Dataset of Table 5.4

	SLP (hPa)	850 hPa GPH (m)	Specific Humidity (g/kg)	Near-Surf. Air Temp. (°C)	Predictor Y
1961	0.504	–0.047	–1.101	–1.042	–1.118
1962	0.549	0.157	–0.458	–0.285	–0.195
1963	0.660	0.276	–0.786	–0.463	0.079
1964	–0.522	0.099	0.826	0.925	1.295
1965	0.389	–0.066	–1.016	–0.782	0.001
1967	–0.436	0.361	1.599	1.517	1.512
1968	2.606	2.048	–0.737	–1.005	–1.004
1969	0.673	1.036	0.375	0.508	0.443
1970	–0.144	–0.823	–1.388	–1.480	–1.495
1971	–1.192	–1.496	–0.809	–0.850	–0.932
1972	0.041	0.215	0.645	0.605	1.154
1973	–0.209	0.369	1.426	0.993	0.694
1974	0.869	0.972	0.534	0.518	0.348
1975	0.036	0.524	1.056	1.163	0.904
1976	0.215	–0.191	–0.696	–0.881	–1.163
1977	–2.247	–2.658	–1.116	–1.274	–1.183
1978	–0.311	–0.552	–0.428	–0.466	–0.650
1979	–0.537	–0.440	0.536	0.443	0.129
1980	0.535	0.559	–0.060	–0.003	0.541
1981	–0.150	–0.932	–1.301	–1.692	–1.558
1984	1.753	1.254	–0.449	–0.257	–0.677
1985	–1.726	–1.829	–0.930	–0.597	–0.976
1986	–0.297	–0.095	0.715	0.432	0.386
1987	–0.697	–0.166	0.421	0.745	0.359
1989	0.666	1.169	0.864	1.044	1.193
1990	–1.027	0.257	2.276	2.183	1.914

For the final predictor estimate for January 1961 is then

$$\hat{y}_{1961}^{org} = -0.942 \times 3.02 + (-5.73) = -8.575$$

The estimate $\hat{y}_i^{org}$ for the observed period is presented in Table 5.4.

d. For the base period and the future period of the CGCM dataset, the mean and standard deviation for the base period dataset are estimated and the standardization is performed with the mean and standard deviation for the base period dataset as in Table 5.6. Also, the standardization for the future period is applied with the mean and standard deviation for the base period, not with the ones for the future period. This is because the statistical behavior might change in the future from climate change compared with the base period dataset. The statistical variation should not be lost due to climate change.

TABLE 5.6
Partial Dataset for Projecting the Future Scenario of the Target Monthly Air Temperature at Cedars, Quebec

Data	Years	SLP (Pa)	850 hPa GPH (m)	Specific Humidity (g/g)	Near-Surf. Air Temp. (°K)	$\hat{y}_i$
Base period	1991	101795.8	1383.3	0.00186	262.784	–9.394
	1992	101705.0	1381.4	0.00159	262.255	–9.412
	1993	101998.7	1438.6	0.00252	269.727	–1.204
	1994	101868.7	1423.3	0.00237	269.243	–1.648
	1995	101954.3	1411.9	0.00174	266.156	–4.471
	1996	101710.5	1402.4	0.00200	267.573	–3.170
	1997	101560.0	1376.6	0.00147	264.103	–6.746
	1998	101680.9	1400.5	0.00258	268.477	–3.306
	1999	101725.0	1386.0	0.00170	264.790	–6.373
	2000	101575.0	1381.2	0.00167	264.896	–6.098
	2001	101263.8	1348.1	0.00181	264.696	–6.963
	2002	101688.0	1385.2	0.00156	264.664	–6.193
	2003	101318.5	1374.6	0.00232	268.076	–3.351
	2004	101739.6	1390.8	0.00182	264.230	–7.267
	2005	101382.3	1363.3	0.00157	265.297	–5.502
	Mean[a]	101680.9	1389.12	0.00185	265.476	–5.730
	Std	279.2	25.20	0.00038	2.723	2.979
Future period	2086	101183.4	1382.9	0.00328	272.198	0.077
	2087	101784.4	1413.0	0.00206	268.762	–1.671
	2088	101544.7	1403.3	0.00275	271.361	0.186
	2089	100895.0	1345.4	0.00261	269.992	–1.666
	2090	101163.3	1397.9	0.00377	276.100	4.261
	2091	101404.4	1387.6	0.00229	269.720	–1.046
	2092	101803.1	1446.6	0.00336	275.881	5.117
	2093	101513.0	1423.5	0.00390	275.490	3.307
	2094	101577.4	1424.6	0.00316	274.392	3.493
	2095	101343.3	1410.2	0.00373	275.595	3.741
	2096	101784.5	1433.7	0.00275	273.529	3.255
	2097	101905.8	1436.7	0.00236	272.298	2.465
	2098	101634.0	1427.1	0.00317	273.130	1.855
	2099	101532.4	1400.4	0.00247	270.491	–0.354
	2100	101637.2	1406.8	0.00229	270.112	–0.439
	Mean[b]	101619.7	1406.8	0.00254	270.404	–0.601
	Std	265.7	22.7	0.00055	2.829	2.775

The dataset of the employed predictors for future projection was extracted from CGCM of Environment Canada and its unit is different from the NCEP/NCAR in Table 5.4.

[a] Mean and standard deviation are estimated with the total dataset of the base period 1850–2005, not only the presented partial dataset.

[b] Mean and standard deviation are estimated with the total dataset of the future period 2006–2100, not only the presented partial dataset.

For example, the SLP value of January 2086 for the future period is 101183.4 Pa and the mean and standard deviation for the base period 1850–2005 are 101680.9 and 279.2, respectively. The standardized value for January 2086 is then (101183.4–101680.9)/279.2 = –1.782. After all the predictor variables are standardized with the mean and standard deviation of the base period for the future period, the future projection of the target predictand can be estimated with $\vec{\mathbf{x}}_i^T \boldsymbol{\beta}$ and the back-standardization should be made with the mean and standard deviation of the observation for the predictand a:

$$\vec{\mathbf{x}}_i^T \boldsymbol{\beta} = \begin{bmatrix} 0 \\ -1.782 \\ -0.247 \\ 3.723 \\ 2.469 \end{bmatrix}^T \times \begin{bmatrix} 1 \\ -0.096 \\ 0.156 \\ -0.278 \\ 1.144 \end{bmatrix} = 1.923$$

$$\hat{y}_{2086}^{org} = 1.923 \times 3.02 + (-5.73) = 0.077$$

Results of multiple linear regression with the CGCM dataset of the base and future periods are presented in Figure 5.5 as well. A steep increase

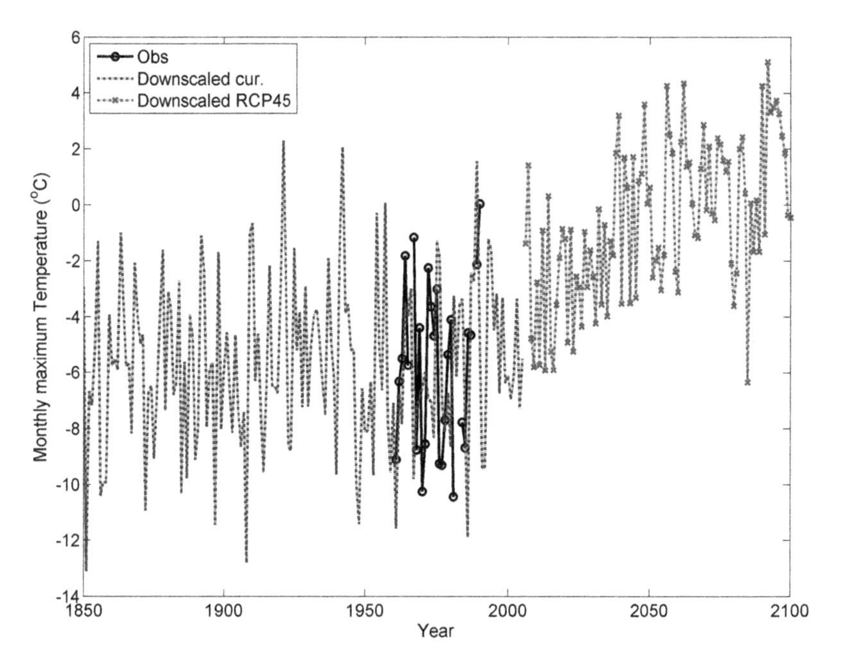

FIGURE 5.5 Downscaled temperature for the base period (dotted line) of 1850–2005 and the future period (dotted line with cross markers) of 2006–2100 employing the near-surface specific humidity of CGCM for the January average as well as the observed (solid line with circles) monthly maximum temperature. See Tables 5.4 and 5.6 for detailed values.

of monthly maximum temperature can be observed for future projection (2006–2100).

Example 5.3

Downscale the future scenario of the January precipitation for Cedars, Quebec, with U (west to east) wind speed and surface air temperature as in the dataset of Table 5.7.

TABLE 5.7
Dataset for Multiple Regression Example of the Target January Precipitation (mm) at Cedars, Quebec, with the Original Dataset (Left-Side) and Standardize Data (Right-Side)

	Org			Standardized (Std.)			
Years	U Wind (m/s)	Surf. Air Temp (°C)	Jan. Precip. (mm)	Std. U Wind	Std. Air Temp	Std. Jan. Precip.	$\hat{y}_i$
1961	7.11	−13.54	39.4	−0.90	−1.04	−0.76	58.69
1962	13.03	−11.34	64.3	1.74	−0.29	0.16	42.11
1963	8.82	−11.86	58.7	−0.14	−0.46	−0.05	57.13
1964	8.84	−7.83	110.7	−0.13	0.92	1.86	69.47
1965	8.98	−12.78	41.1	−0.07	−0.78	−0.70	53.64
1967	10.9	−6.11	84.9	0.79	1.52	0.91	66.65
1968	7.24	−13.43	56.0	−0.84	−1.00	−0.15	58.51
1969	7.05	−9.04	107.8	−0.93	0.51	1.76	72.80
1970	8.08	−14.81	31.9	−0.47	−1.48	−1.04	50.95
1971	10.09	−12.98	86.4	0.43	−0.85	0.97	48.65
1972	13.18	−8.76	36.3	1.80	0.61	−0.87	49.49
1973	10.84	−7.63	30.0	0.76	0.99	−1.11	62.19
1974	12.85	−9.01	30.5	1.66	0.52	−1.09	50.01
1975	10.67	−7.14	47.3	0.69	1.16	−0.47	64.39
1976	9.11	−13.07	84.8	−0.01	−0.88	0.91	52.24
1977	8.6	−14.21	86.1	−0.24	−1.27	0.96	50.74
1978	5.73	−11.87	96.5	−1.52	−0.47	1.34	69.29
1979	5.21	−9.23	54.2	−1.75	0.44	−0.21	79.48
1980	9.41	−10.52	19.8	0.12	0.00	−1.48	58.92
1981	8.16	−15.43	19.0	−0.43	−1.69	−1.51	48.73
1984	7.93	−11.26	44.4	−0.54	−0.26	−0.58	62.48
1985	8.17	−12.25	40.0	−0.43	−0.60	−0.74	58.49
1986	9.35	−9.26	96.4	0.10	0.43	1.34	63.05
1987	5.12	−8.35	72.7	−1.79	0.75	0.47	82.55
1989	11.36	−7.48	53.6	0.99	1.04	−0.24	60.60
1990	11.6	−4.18	68.3	1.10	2.18	0.30	69.85
Mean	9.1	−10.5	60.04				
Std	2.24	2.90	27.18				

Its observational dataset from NCAR/NCEP for U wind speed and surface air temperature and from the weather station for January precipitation is presented in Table 5.7. The relationships between the January precipitation and the two predictors (U wind and air temperature) are shown in Figure 5.6. Note that the U wind has a negative relation to the predictand (January precipitation), while the surface air temperature has a positive relation. These correlations are not much stronger in the case of the previous example for the January temperature. It is natural that a precipitation variable is not strongly related with other climate variables and its correlation is weak. As mentioned, the standardization to avoid the scale difference is applied, as shown on the right side of Table 5.7.

Solution:

Multiple linear regression model is fitted to the standardized data, and its parameter vector is estimated with Eq. (5.12) for this example as

$$\hat{\beta} = [0, -0.326, 0.329]^T$$

The predictor estimate is then made by applying Eq. (5.11), ignoring the noise term as

$$Y_i = \vec{\mathbf{x}}_i^T \beta \qquad (5.16)$$

The value for January 1961 is estimated as

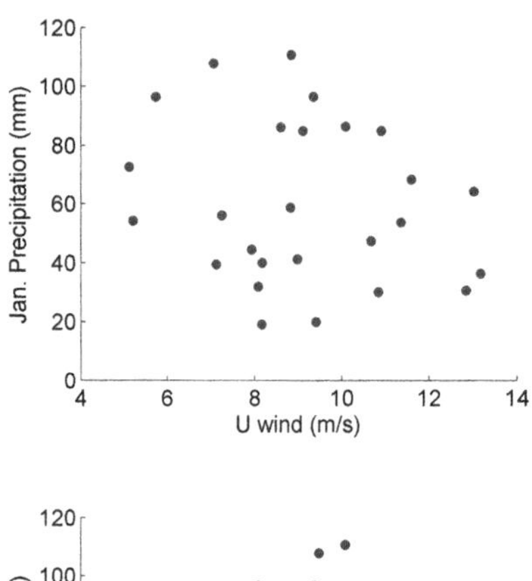

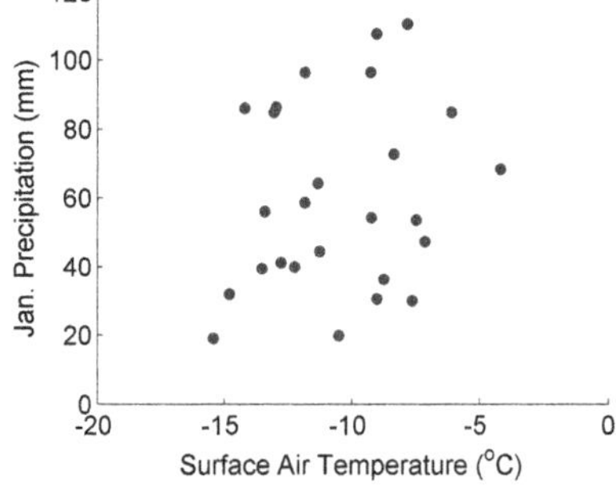

FIGURE 5.6 Scatterplot between the target monthly maximum temperature (°C) at Cedars, Quebec, and the four atmospheric predictors obtained from NCEP/NCAR reanalysis data with multiple linear regression. Note that only the January data were employed to avoid the seasonality of the dataset.

$$[1,-0.9,-1.04]\times[0,-0.326,0.329]^T = -0.0488$$

Here, the vector multiplication is applied.

The predictor estimate for January 1961 is then

$$\hat{y}_{1961}^{org} = -0.0488\times 27.18+(60.04) = 58.69$$

All the estimate $\hat{y}_i$ for the observed period is presented in Table 5.7.

For the base and future periods of the CGCM dataset, as mentioned in the previous example, the mean and standard deviation for the base period dataset are estimated, and the standardization is performed with the mean and standard deviation for the base period dataset as well as for the future, and not with the ones for the future period. In this case, the statistical variation through time (called nonstationarity) due to climate change should not be lost.

The results of multiple linear regression with the CGCM dataset of the base and future periods for the January precipitation are presented in Figure 5.7 as well as Table 5.8. The figure presents that there is a slight increase of January precipitation for both RCP45 and RCP85 cases with not much difference in magnitudes.

Note that the variability of the downscaled January precipitation is much lower than the observed data. This variance deflation results from low correlation between the predictors and predictand as well as nonconsideration of the error variability. In other words, ε_i is set to be zero all the time and its variance is zero, while the variance is σ_ε^2. This variance increases when its prediction is not accurate, such as the case of the precipitation predictand with low correlation, with the predictors presented in Figure 5.7.

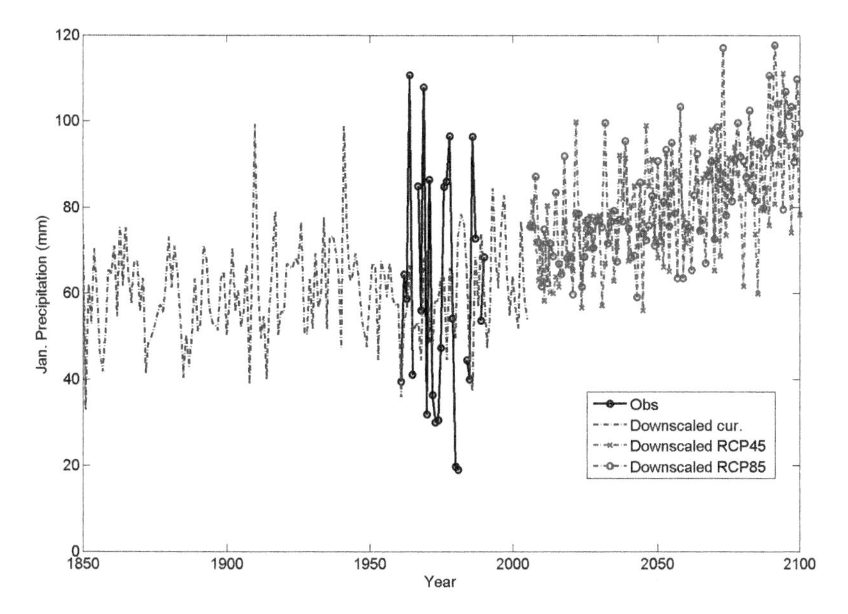

FIGURE 5.7 (See color insert.) Downscaled January precipitation for the base period (dotted line) of 1850–2005 and the future period (dotted line with cross and circle markers for RCP4.5 and RCP8.5, respectively) of 2006–2100 employing the U wind speed and near-surface temperature of CGCM for the January average as well as the observed January precipitation (solid line with circles). See Tables 5.7 and 5.8 for detailed values.

TABLE 5.8
Partial Dataset for Projecting the Future Scenario of the Target January Precipitation at Cedars, Quebec, for the Future Period of RCP45 (Left Side) and RCP85 (Right Side)*

	RCP45			RCP45		
Years	**U Wind (m/s)**	**Air Temp (°K)**	$\hat{y}_i$	**U Wind (m/s)**	**Air Temp (°K)**	$\hat{y}_i$
2091	6.574	269.720	80.697	4.217	278.295	117.701
2092	9.561	275.881	89.784	9.441	280.075	104.019
2093	7.029	275.490	97.965	8.151	276.484	97.039
2094	2.560	274.392	111.069	11.352	274.787	79.490
2095	7.846	275.595	95.257	5.058	275.955	106.867
2096	6.250	273.529	94.431	6.249	275.588	101.206
2097	10.622	272.298	74.036	8.985	279.354	103.355
2098	6.657	273.130	91.597	7.846	274.152	90.509
2099	3.088	270.491	96.268	3.492	275.048	109.740
2100	7.580	270.112	78.225	3.231	270.972	97.316

* The dataset of the predictors employed for future projection was extracted from CGCM in Environment Canada and its unit is different from the NCEP/NCAR in Table 5.7.

To remedy this deflation, one can simulate the error term with the estimated standard deviation such as Eq. (5.8). This simulation is called "stochastic simulation."

5.2 PREDICTOR SELECTION

There is a large set of candidate predictor variables (a number of climate variables from GCM outputs) for the multiple linear regression downscaling. The remaining important argument is which predictors are best, since unnecessary predictors will add noise to the estimation. The question of selecting a small subset from candidate predictor variables that still has good predictive ability must be answered. The predictor selection involves an attempt to find the best model and to limit the number of predictor variables among a large set of possible candidates. Among others, one of the most widely known techniques of predictor selection is stepwise regression (SWR). Efroymson (1966) proposed the idea of SWR, and Draper and Smith (1966) developed the SWR. A possible alternative of the SWR was suggested by Tibshirani (1996) as Least Absolute Shrinkage and Selection Operator (LASSO). This technique is also explained.

5.2.1 Stepwise Regression

The procedure of SWR is as follows.

1. Set a significance level that decides to enter a predictor into the stepwise model (alpha-to-enter, α_E) or to remove a predictor from the stepwise model (alpha-to-remove, α_M). The significance level can be different for entering

and removing, but normally the same in practice. Higher than the usual 0.05 level is employed typically as $\alpha_E = \alpha_M = 0.15$, so that entering and removing a predictor are not too difficult.

2. Start the SWR with no predictor.
   (2.1) Fit each of the one-predictor models as $Y = \beta_0 + \beta_1 x^1$, $Y = \beta_0 + \beta_2 x^2$, … , and $Y = \beta_0 + \beta_S x^S$, where S is the number of variables of interest. Note that the noise term is eliminated for simplicity unlike Eq. (5.1).
   (2.2) Select the first predictor that has the smallest p-value among the ones whose p-value is less than α_E for the null hypothesis test of $H_0 : \beta_j = 0$, $j = 1, \ldots, S$. If no predictor has a p-value less than α_E, then stop the procedure
3. Suppose that x^3 has the smallest p-value and less than α_E.
   (3.1) Fit each of the two predictor models including the first predictor x^3 as $Y = \beta_0 + \beta_3 x^3 + \beta_1 x^1$, $Y = \beta_0 + \beta_3 x^3 + \beta_2 x^2, \ldots,$ and $Y = \beta_0 + \beta_3 x^3 + \beta_S x^S$.
   (3.2) Add the predictor that has the smallest p-value with the t-test among the variables whose p-value is less than α_E.
   (3.3) Stop the procedure if no predictor has a p-value less than α_E. Then, the one-predictor model becomes the final model as $Y = \beta_0 + \beta_3 x^3$.
4. Suppose that the second variable x^2 is chosen as the best second predictor and it is entered into the stepwise model (i.e., $Y = \beta_0 + \beta_3 x^3 + \beta_2 x^2$).
   (4.1) Step back and see if entering x^2 affects the significance of the x^3 predictor by checking the p-value of the t-test. If the p-value of β_3 is greater than α_M, remove the x^3 predictor from the stepwise model.
   (4.2) Otherwise, keep the stepwise model as is.
5. Continue the steps until no predictor is added.

Example 5.4

Select critical variables to fit a multiple linear regression model to the monthly average of daily maximum temperature at Cedars in January by employing stepwise predictor selection among the variables in Table 5.9.

Solution:

The dataset is presented in Table 5.9. The target variable is the same as in Example 5.1, the maximum temperature of Cedars. Here the variables employed are SLP and 500 and 850 hPa GPHs (GPH_{850}, GPH_{500}) of the January average for X_1, X_2, and X_3 variables. At first, the data must be standardized as shown at the right side of Table 5.9 to make the scale similar to each other. Otherwise, an irrelevant variable with high variability might take on high influence in this regression. With this standardized dataset, the procedure described earlier is used:

1. Set the significance level of entrance and removal as 0.15 ($\alpha_E = \alpha_M = 0.15$).
2. Start with no predictor:
   (2.1) Fit each of the predictors for the three variables as presented in the simple linear regression model. For example, the estimated parameter for the first SLP predictor is $\beta_3 = 0.725$. All the estimated parameters are presented in Table 5.10. Note that $\beta_0 = 0$

TABLE 5.9
Dataset for Stepwise Predictor Selection of the Target Monthly Air Temperature at Cedars, Quebec, (Y) with the Original Dataset (Left-Side) and Standardized Data (Right-Side)

	Org				Standardized Data			
Years	SLP (hPa)	850 hPa GPH (gpm)	500 hPa GPH (gpm)	Y	SLP	850 hPa GPH	500 hPa GPH	Y
1961	1017.1	1372.4	5273.6	−9.106	0.504	−0.047	−0.952	−1.118
1962	1017.2	1377.8	5324.9	−6.319	0.549	0.157	−0.045	−0.195
1963	1017.5	1381.0	5313.0	−5.490	0.660	0.276	−0.254	0.079
1964	1014.1	1376.3	5364.5	−1.819	−0.522	0.099	0.657	1.295
1965	1016.7	1371.9	5309.5	−5.726	0.389	−0.066	−0.317	0.001
1967	1014.4	1383.2	5373.5	−1.165	−0.436	0.361	0.816	1.512
1968	1023.1	1428.1	5404.6	−8.761	2.606	2.048	1.365	−1.004
1969	1017.6	1401.2	5375.8	−4.394	0.673	1.036	0.857	0.443
1970	1015.2	1351.8	5250.0	−10.245	−0.144	−0.823	−1.369	−1.495
1971	1012.2	1333.9	5239.5	−8.545	−1.192	−1.496	−1.554	−0.932
1972	1015.7	1379.4	5334.3	−2.245	0.041	0.215	0.121	1.154
1973	1015.0	1383.5	5364.7	−3.635	−0.209	0.369	0.659	0.694
1974	1018.1	1399.5	5392.4	−4.681	0.869	0.972	1.150	0.348
1975	1015.7	1387.5	5371.2	−3.000	0.036	0.524	0.775	0.904
1976	1016.2	1368.5	5309.1	−9.242	0.215	−0.191	−0.323	−1.163
1977	1009.2	1303.0	5190.4	−9.303	−2.247	−2.658	−2.423	−1.183
1978	1014.7	1359.0	5320.2	−7.694	−0.311	−0.552	−0.127	−0.650
1979	1014.1	1361.9	5338.9	−5.342	−0.537	−0.440	0.204	0.129
1980	1017.2	1388.5	5342.6	−4.097	0.535	0.559	0.269	0.541
1981	1015.2	1348.9	5252.8	−10.435	−0.150	−0.932	−1.320	−1.558
1984	1020.6	1407.0	5349.5	−7.774	1.753	1.254	0.391	−0.677
1985	1010.7	1325.0	5240.7	−8.677	−1.726	−1.829	−1.533	−0.976
1986	1014.8	1371.1	5334.1	−4.565	−0.297	−0.095	0.118	0.386
1987	1013.6	1369.2	5347.3	−4.645	−0.697	−0.166	0.352	0.359
1989	1017.5	1404.7	5398.7	−2.129	0.666	1.169	1.261	1.193
1990	1012.7	1380.5	5396.5	0.048	−1.027	0.257	1.222	1.914
Mean	1015.6	1373.6	5327.4	−5.73	0.00	0.00	0.00	0.00
Std	2.87	26.57	56.54	3.02	1.00	1.00	1.00	1.00

always, because the variables are standardized to zero mean. Also, T-statistic and its corresponding p-value are estimated. For example, T-statistic for X_3 variable is $T = (0.725-0)/0.14 = 5.164$ [see Eq. (5.9) for detail] and p-value as 2.74×10^{-5} with $n-2$ $(26-2 = 24)$ degrees of freedom for the t-distribution.

(2.2) As shown in Table 5.10, select one variable with the smallest p-value indicating the strongest linear relation with the predictand. Here, the X_3 variable can be selected with $2.74 \times 10^{-5} \ll 0.15 (= \alpha E)$.

TABLE 5.10
Example of Stepwise Predictor Selection for Three Variables of X_1 (SLP), X_2 (850 hPa GPH), and X_3 (500 hPa GPH)

	Variable	β_i	SE	*t*-stat	*p*-val
Initiate	X_1	–0.027	0.204	–0.134	0.895
	X_2	0.448	0.182	2.457	0.022
	X_3	0.725	0.140	5.164	2.7E-05
X_3 Selection	X_1	–0.620	0.114	–5.439	1.6E-05
	X_2	–0.904	0.245	–3.685	0.001
	X_3	0.725	0.140	5.164	2.7E-05
X_1 Selection	X_1	–0.620	0.114	–5.439	1.6E-05
	X_2	2.015	0.598	3.373	0.003
	X_3	1.069	0.114	9.377	2.5E-09
X_2 Selection	X_1	–1.697	0.333	–5.097	4.2E-05
	X_2	2.015	0.598	3.373	0.003
	X_3	–0.119	0.365	–0.326	0.747
X_3 Removal	X_1	–1.601	0.152	–10.531	2.9E-10
	X_2	1.827	0.152	12.019	2.1E-11
	X_3	–0.119	0.365	–0.326	0.747

3. Test two predictor models with the condition including X_3 variable always:
   (3.1) Estimate the parameters and *T*-statistic as well as *p*-value. For example, the estimated parameter for the X_2 variable, GPH_{850} predictor is $\beta_2 = -0.904$ in the model of $Y = \beta_0 + \beta_3 x^3 + \beta_2 x^2$ as presented in Table 5.10. From the estimated *p*-values for X_2 and X_1 variables, it can be noted that it is feasible to include the X_1 variable as a predictor.
   (3.2) Check whether adding the X_1 variable affects the significance of the X_3 variable. In the X_1 selection row, the *p*-value of X_3 is $2.5 \times 10^{-9} << 0.15$, indicating that the inclusion of the X_1 variable does not affect the significance of X_3.
4. Test three parameter models with the condition of including the X_1 and X_3 variables:
   (4.1) Estimate the parameter for the X_2 variable ($\beta_2 = 2.015$) and *T*-statistic ($T = 3.373$) as well as the *p*-value [0.003, which is <0.15 (= αE)]. Therefore, X_2 can be added to the model.
   (4.2) Check whether adding the X_2 variable affects the significance of the X_1 and X_3 variables. In the X_2 selection row, the *p*-values of X_1 and X_3 are 4.2×10^{-5} (<<0.15) and 0.747 (>0.15), respectively. The values indicate that the inclusion of the X_2 variable affects the significance of X_3.
   (4.3) Remove the X_3 variable from the model.
5. Check the *p*-values for the other variables whether there is any variable to enter or remove. The *p*-values in the X_3 removal row shows that no variables need to be entered or removed.
6. Stop the procedure. The final model is $Y = \beta_0 + \beta_1 x^1 + \beta_2 x^2 = \beta_1 x^1 + \beta_2 x^2$ with $\beta_1 = -1.601$, $\beta_2 = 1.827$, and $\beta_0 = 0$.

5.2.2 Least Absolute Shrinkage and Selection Operator

The parameters of a regression model can be calculated by minimizing the cost function defined as

$$\text{Total cost} = \sum_{i=1}^{n} (y_i - \hat{y}_i)^2 = \sum_{i=1}^{n} \varepsilon^2 \tag{5.17}$$

where $\hat{y}_i$ is the regression estimate of y_i for $i = 1, \ldots, n$ and n is the number of data. However, this definition of the cost function has some disadvantages. As model complexity increases, using this method may lead to a model with low bias and high variance. It means that although the model may fit well the training set (low bias), it predicts the new test points very poorly (high variance). For example, in a polynomial model, as the complexity (degree of polynomial) increases, the model performance for the training set improves but the values of parameters grow, and the learned function starts to oscillate. This behavior is not desirable for a model, since it makes it sensitive to a new random test point or small fluctuations in the training set. Therefore, always a tradeoff between bias and variance is needed and should be considered in the learning process.

Predictor selection, also known as feature selection, is a powerful tool to select the best subset of relevant features for estimating the regression model. To address these issues, the regularization methods are developed. Regularization is a process of introducing additional information to solve a feature selection problem or to prevent overfitting by penalizing high-valued regression coefficients. Generally speaking, regularization techniques can be expressed as

$$\begin{aligned} \hat{\boldsymbol{\beta}} &= \underset{\beta \in \Re^S}{\arg\min} \left\| \mathbf{y} - \mathbf{x}\boldsymbol{\beta} \right\|_2^2 + \lambda R(\boldsymbol{\beta}) \\ \left\| \mathbf{y} - \mathbf{x}\boldsymbol{\beta} \right\|_2^2 &= \sum_{i=1}^{n} (y_i - \hat{y}_i)^2 \end{aligned} \tag{5.18}$$

where (1) S is the number of predictors or features in the model (i.e., $\boldsymbol{\beta} = [\beta_1, \ldots, \beta_S]$); (2) λ is a tuning parameter controlling the regularization ability of the model; and (3) the term R is known as a penalty or regularizer. Applying regularization means modifying the regression problem, such as overfitting.

The LASSO regression (Tibshirani, 1996) is one of the regularization techniques. In this method, the penalty term is defined as L1 norm of $\boldsymbol{\beta}$ as

$$R(\boldsymbol{\beta}) = \left\| \boldsymbol{\beta} \right\|_1 = \sum_{j=1}^{S} \left| \beta_j \right| \tag{5.19}$$

Therefore, the penalized form of the total cost function can be written as

$$\text{Total cost} = \sum_{i=1}^{n} (y_i - \hat{y}_i)^2 + \lambda \sum_{j=1}^{S} \left| \beta_j \right| \tag{5.20}$$

By imposing an L1 penalty to the total cost, the growth of the value of model parameters can be controlled, so a model with lower variance would be yielded. Furthermore, the nature of penalty term in the LASSO regression provides the chance for setting some parameters exactly to zero, i.e., performing predictor selection.

As λ increases (i.e., higher weight on the penalty term), more parameters are set to zero. Therefore, the predictor selection is automatically obtained. Choosing an appropriate value of λ is a key step in the LASSO regression. Sometimes, it is necessary to find a model that has only p predictors i.e., only p nonzero parameters in $\hat{\boldsymbol{\beta}}$. In this case, the following algorithm is proposed:

1. Split the dataset randomly into two subsets: a training set and a validation set. There is no specific rule for the size of each subset, but typically they are 90%–10% or 80%–20% of an original dataset.
2. Choose a range of m different values of tuning parameters $\{\lambda_1, \lambda_2, \ldots, \lambda_m\}$. The goal is to find the best value of λ among the m elements of the λ values. The number and values of λ, chosen for evaluation, are determined, based on experience. If one cannot estimate a proper range of λ, a wide range should be studied. It should consist of small to large values of tuning parameters.
3. Fit a model to the training set using the LASSO regression for each specific value of tuning parameters (the fitting algorithm is explained in detail in the following).
4. Select those tuning parameters whose models have only p nonzero parameters.
5. Evaluate the error of selected model for the validation set. The model with minimum validation error would be selected as the desired model.

When the number of observations in the dataset is limited, the K-fold cross validation (K-CV) method would be helpful instead of splitting the dataset. The K-CV procedure can be described as follows:

1. Divide the dataset randomly into K-subsets of roughly equal size, $F_1, F_2, \ldots, F_K$. The value of K depends on the size of the dataset. For example, in a dataset with fewer number of observations, choosing a small value of K leads to a poor prediction, since the size of training set in each step would be too small. Typically, K is 5 or 10. When K is equal to same as the number of data n, one data value is left out at each step, called leave-one-out cross validation method. However, because of its high computational load, this method of cross validation is not applicable in many cases.
2. Choose m different values of tuning parameters $\{\lambda_1, \lambda_2, \ldots, \lambda_m\}$ to study.
3. For each specific value of tuning parameters $\lambda_1, \lambda_2, \ldots, \lambda_m$:
   (3.1) Consider F_j as a validation set and the rest of data as a training set.
   (3.2) Fit the regression model on the training set (the fitting algorithm is explained in detail in the following) and evaluate it for the validation set. The validation error is defined as

$$e_j(\lambda) = \sum_{i \in F_j} (y_i - \hat{y}_i^{-j}(\lambda))^2 \quad (5.21)$$

in which, $\hat{y}_i^{-j}(\lambda)$ is the ith estimate with the fitted model for the training set, excluding the validation set F_j for a given value of λ.

(3.3) Repeat steps (3.1) and (3.2) for all the folds ($j = 1, 2, \ldots, K$). So, all of the subsets are applied one time as the validation set and K-1 times as the training set.

(3.4) Compute the mean-squared error (MSE), which is the average of validation errors over all K-folds as

$$MSE = \frac{1}{n}\sum_{j=1}^{K} e_j(\lambda) = \frac{1}{n}\sum_{j=1}^{K}\sum_{i \in F_j} (y_i - \hat{y}_i^{-j}(\lambda))^2 \quad (5.22)$$

where n is the size of the dataset.

4. Select the tuning parameter with a minimum MSE as the best one.

To fit a model with the LASSO regression for the given λ value, the total cost function should be minimized. Because the derivative of $|\boldsymbol{\beta}|$ does not exist, there is no closed-form solution for this model. Instead, the coordinate descent algorithm can be applied. It is a repetitive process that, at each iteration, all parameters are fixed, except the ith parameter. β_i is found in a way that optimizes the total cost function. The total cost can be expressed as

$$\begin{aligned} \text{Total cost} &= SSE + \lambda \sum_{i=1}^{S} |\beta_i| \\ &= \sum_{j=1}^{n} \left(y_j - \sum_{i=1}^{S} \beta_i x_{ji} \right)^2 + \lambda \sum_{i=1}^{S} |\beta_i| \end{aligned} \quad (5.23)$$

where $\mathbf{X}$ (a matrix with n rows and S columns) is the feature matrix consisting of standardized data. Here, x_{ji} is the element of feature matrix in the jth row and ith column. For the sake of simplicity, the derivative of SSE with respect to β_i is calculated separately, and then the derivative of L1 norm of $\boldsymbol{\beta}$ is studied.

$$\begin{aligned} \frac{\partial}{\partial \beta_i}(SSE) &= -2\sum_{j=1}^{n} x_{ji} \left(y_j - \sum_{k=1}^{S} \beta_k x_{jk} \right) \\ &= -2\sum_{j=1}^{n} x_{ji} \left(y_j - \sum_{k \neq i} \beta_k x_{jk} - \beta_i x_{ji} \right) \\ &= -2\sum_{j=1}^{n} x_{ji} \left(y_j - \sum_{k \neq i} \beta_k x_{jk} \right) + 2\beta_i \sum_{j=1}^{n} x_{ji}^2 \end{aligned} \quad (5.24)$$

Substituting $\sum_{j=1}^{n} x_{ji}^2$ with z_i, the earlier equation can be written as

$$\frac{\partial}{\partial \beta_i}(SSE) = -2\sum_{j=1}^{n} x_{ji}\left(y_j - \sum_{k \neq i} \beta_k x_{jk}\right) + 2\beta_i z_i \tag{5.25}$$

For the penalty part, we have to deal with the subgradient of L1 term, which is defined as

$$\frac{\partial}{\partial \beta_i}\left(\lambda \sum_{i=1}^{p} |\beta_i|\right) = \frac{\partial}{\partial \beta_i}(\lambda|\beta_i|) = \begin{cases} -\lambda & \beta_i < 0 \\ (-\lambda, \lambda) & \beta_i = 0 \\ \lambda & 0 < \beta_i \end{cases} \tag{5.26}$$

Putting the equations all together as

$$\begin{aligned} \frac{\partial}{\partial \beta_i}(\text{Total cost}) &= \frac{\partial}{\partial \beta_i}(SSE) + \frac{\partial}{\partial \beta_i}\left(\lambda \sum_{i=1}^{s} |\beta_i|\right) \\ &= -2\rho_i + 2\beta_i z_i + \begin{cases} -\lambda & \beta_i < 0 \\ (-\lambda, \lambda) & \beta_i = 0 \\ \lambda & 0 < \beta_i \end{cases} \\ &= \begin{cases} -2\rho_i + 2\beta_i z_i - \lambda & \beta_i < 0 \\ (-2\rho_i - \lambda, -2\rho_i + \lambda) & \beta_i = 0 \\ -2\rho_i + 2\beta_i z_i + \lambda & 0 < \beta_i \end{cases} \end{aligned} \tag{5.27}$$

where

$$\rho_i = \sum_{j=1}^{n} x_{ji}\left(y_j - \sum_{k \neq i} \beta_k x_{jk}\right) = \mathbf{x}_i * \left(\mathbf{y} - \hat{\mathbf{y}} + \beta_i \mathbf{x}_i\right)$$

Since the goal is one of minimizing the total cost function, the derivative $\frac{\partial}{\partial \beta_i}(\text{Total cost})$ of total cost function must be taken, and consequently, the subgradient should be set equal to zero. Finally, the desired β_i that optimizes the total cost function would be obtained as

$$\beta_i = \begin{cases} \dfrac{\rho_i + \dfrac{\lambda}{2}}{z_i} & \rho_i < -\dfrac{\lambda}{2} \\ 0 & -\dfrac{\lambda}{2} \leq \rho_i \leq \dfrac{\lambda}{2} \\ \dfrac{\rho_i - \dfrac{\lambda}{2}}{z_i} & \rho_i > \dfrac{\lambda}{2} \end{cases} \tag{5.28}$$

In a nut shell, the steps of coordinate descent algorithm can be written as follows:

1. Create a feature matrix and response vector. The response vector (**y**) is a column vector with S elements containing the standardized values of responses.
2. Assign a set of initial parameters ($\boldsymbol{\beta}$). It can be a vector of zeros or any other smart guess.
3. Regularize and update parameters. For this purpose, start with the first element of parameter vector, i.e., β_1. Go through the following steps and repeat the steps for the second parameter, and so on for all of $i = 1, 2, \ldots, S$:
   (3.1) Compute the prediction vector $\hat{\mathbf{y}}$, and the terms ρ_i and z_i:

$$\hat{\mathbf{y}} = \mathbf{X} * \hat{\boldsymbol{\beta}}$$

$$\rho_i = \sum_{j=1}^{n} x_{ji}\left(y_j - \sum_{k \neq i} \beta_k x_{jk}\right) = \mathbf{x}_i * (\mathbf{y} - \hat{\mathbf{y}} + \beta_i \mathbf{x}_i)$$

$$z_i = \sum_{j=1}^{n} x_{ji}^2 = \mathbf{x}_i^T \mathbf{x}_i$$

   where $\mathbf{x}_i$ is the ith column of feature matrix.
   (3.2) Regularize the ith parameter as in Eq. (5.28):

$$\beta_i(regularized) = \begin{cases} \dfrac{\rho_i + \dfrac{\lambda}{2}}{z_i} & \rho_i < -\dfrac{\lambda}{2} \\ 0 & -\dfrac{\lambda}{2} \leq \rho_i \leq \dfrac{\lambda}{2} \\ \dfrac{\rho_i - \dfrac{\lambda}{2}}{z_i} & \rho_i > \dfrac{\lambda}{2} \end{cases}$$

   (3.3) Calculate the change of the ith parameter and then update it:

$$c_i = \left|\beta_i(regularized) - \beta_i\right|$$

$$\beta_i = \beta_i(regularized)$$

   where **c** is the change vector with elements of c_i, $i = 1, \ldots, S$.
4. Stop if max (c_i, $i = 1, \ldots, S$) < ξ; otherwise, repeat step (3) for an updated parameter vector. The value of tolerance (ξ) is chosen with a very low value to avoid that the process is stopped before convergence such as 10^{-4}. Note that the standardized dataset is applied, and as a result, the intercept is equal to zero (say, $\beta_0 = 0$). Finally, the vector of estimated parameters is

$$\hat{\beta} = \begin{bmatrix} \hat{\beta}_1 & \hat{\beta}_1 & \cdots & \hat{\beta}_S \end{bmatrix}$$

Example 5.5

Select predictor variables by employing LASSO for the prediction of maximum temperature based on four predictors: 500 hPa wind speed (x_1), 500 hPa geopotential (x_2), 850 hPa geopotential (x_3), and 1,000 hPa specific humidity (x_4) (Table 5.4).

Solution:

The problem is solved with two different approaches. The first approach emphasizes the predictor selection. In this approach, a model with two predictors is interesting (i.e., the number of predictors is preassigned. The second approach applies LASSO to prevent overfitting.

First approach:

1. Standardize the data. Shuffle the standardized dataset randomly as shown in Table 5.11. Select 90% of shuffled data, i.e., the first 23 data as the training set. The rest of the data is applied as the validation set.
2. Consider 100 different values of λ in [0.002, 4] (i.e., λ = [0.0020, 0.0434, ..., 3.9592, 4.000].)
3. Create feature matrix and response vector based on the data in the training set.

$$\mathbf{X} = \begin{bmatrix} 0.5052 & -0.0467 & -1.1000 & -1.0441 \\ 0.5505 & 0.1573 & 0.4574 & -0.2866 \\ \vdots & \vdots & \vdots & \vdots \\ -1.1882 & -1.4964 & -0.8085 & -0.8524 \end{bmatrix}$$

$$\mathbf{y} = \begin{bmatrix} -1.1179 \\ -0.1950 \\ \vdots \\ -0.9321 \end{bmatrix}$$

4. Start with the first value of λ (0.0020). Fit a model for this specific value of λ to the training set with the gradient descent algorithm. For this purpose, set initial parameters as a vector of zeros (i.e., $\boldsymbol{\beta}$ = [0, 0, 0, 0]) and tolerance (ξ) as 10^{-4}, then follow the procedure given later:

(4.1) Start with the first parameter. For this purpose, calculate the prediction vector, and the terms ρ_1 and z_1 corresponding to this parameter.

Since the initial parameters are equal to zero, the prediction vector in this step is also a vector of zeros, i.e., $\hat{\mathbf{y}} = \begin{bmatrix} 0 & 0 & \cdots & 0 \end{bmatrix}^T$

$$\rho_1 = \begin{bmatrix} 0.5052 \\ 0.5505 \\ \vdots \\ -1.1882 \end{bmatrix}^T \times \left[\begin{bmatrix} -1.1179 \\ -0.195 \\ \vdots \\ -0.9321 \end{bmatrix} - \begin{bmatrix} 0 \\ 0 \\ \vdots \\ 0 \end{bmatrix} + \left[0 \times \begin{bmatrix} 0.5052 \\ 0.5505 \\ \vdots \\ -1.1882 \end{bmatrix} \right] \right]$$

$$= -0.0787$$

TABLE 5.11
Randomly Shuffled Standardized Data for the LASSO Example

	x_1	x_2	x_3	x_4	y
1	0.5052	–0.0467	–1.1000	–1.0441	–1.1179
2	0.5505	0.1573	0.4574	–0.2866	–0.1950
3	0.6585	0.2763	–0.7581	–0.4645	0.0795
4	–2.2439	–2.6583	–1.1149	–1.2759	–1.1831
5	0.6725	1.0361	0.3745	0.5066	0.4424
6	–1.7247	–1.8291	–0.9298	–0.5990	–0.9758
7	–0.2962	–0.0952	0.7149	–0.4310	0.3858
8	–1.0244	0.2567	2.2745	2.1838	1.1932
9	0.8676	0.9718	0.5340	0.5172	0.3474
10	–0.4355	0.3613	1.5979	1.5166	1.5116
11	2.6028	2.0486	–0.7362	–1.0069	–1.0036
12	0.0383	0.5239	1.0553	1.1631	0.9040
13	1.7526	1.2548	–0.4489	–0.2590	–0.6768
14	–0.1429	–0.8227	–1.3872	–1.4824	–1.4950
15	–0.2091	0.3696	1.4255	0.9924	0.6937
16	0.5366	0.5589	–0.0596	–0.0045	0.5407
17	0.6655	1.1697	0.8638	1.0431	1.1924
18	–0.5226	0.0990	0.8255	0.9241	1.2950
19	–0.1498	–0.9319	–1.3000	–1.6925	–1.5579
20	0.2160	–0.1912	–0.6957	–0.8834	–1.1629
21	0.0418	0.2153	0.6447	0.6045	1.1540
22	–0.6969	0.1656	0.4213	0.7448	0.3593
23	–1.1882	–1.4964	–0.8085	–0.8524	–0.9321
24	0.3902	–0.0662	–1.0149	–0.7838	0.0013
25	–0.5366	–0.4400	0.5362	0.4417	0.1285
26	–0.3101	–0.5518	–0.4277	–0.4676	–0.6503

$$z_1 = \begin{bmatrix} 0.5052 \\ 0.5505 \\ \vdots \\ -1.1882 \end{bmatrix}^T \times \begin{bmatrix} 0.5052 \\ 0.5505 \\ \vdots \\ -1.1882 \end{bmatrix} = 24.4174$$

(4.2) Regularize the first parameter. Since $\rho_1 = -0.0787$ is less than $-\frac{\lambda}{2} = -0.0010$, the regularized β_1 is equal to

$$\frac{\rho_1 + \frac{\lambda}{2}}{z_1} = \frac{-0.0787 + 0.0010}{24.4174} = -0.0032.$$

(4.3) Calculate the change of the first parameter and then update the vector of parameters:

$$c_1 = |0 - (-0.0032)| = 0.0032$$

$$\hat{\beta} = [-0.0032, 0, 0, 0]$$

(4.4) Repeat these steps for the second parameter and so on, as shown in Table 5.12. Use the updated vector of parameters at each iteration. Since the maximum change is greater than the tolerance value (i.e., $c_3 > \xi$, $0.6803 > 10^{-4}$), the steps should be repeated again and again until the maximum change falls below the tolerance. The final vector of parameter for $\lambda = 0.0020$ would be

$$\hat{\beta} = [0.4943, -0.4920, -0.1361, 1.2909].$$

5. Repeat the same process for the second tuning parameter and so on. The vector parameter for different values of tuning parameters is shown in Table 5.13.
6. Select those estimations of model that have only two predictors. These estimations are used for more evaluations.

TABLE 5.12
First Iteration of Coordinate Descent Algorithm for $\lambda = 0.0032$

	$\hat{\beta}$	$\mathbf{x}_i$	$\hat{\mathbf{Y}}$	ρ_i	z_i	β_i	c_i	$\hat{\beta}_{updated}$
Initial condition	$\begin{bmatrix}0\\0\\0\\0\end{bmatrix}$							
$i = 1$	$\begin{bmatrix}0\\0\\0\\0\end{bmatrix}$	$\begin{bmatrix}0.5052\\0.5505\\\vdots\\-1.1882\end{bmatrix}$	$\begin{bmatrix}0\\0\\\vdots\\0\end{bmatrix}$	−0.0787	24.4174	−0.0032	0.0032	$\begin{bmatrix}-0.0032\\0\\0\\0\end{bmatrix}$
$i = 2$	$\begin{bmatrix}-0.0032\\0\\0\\0\end{bmatrix}$	$\begin{bmatrix}-0.0467\\0.1573\\\vdots\\-1.4964\end{bmatrix}$	$\begin{bmatrix}-0.0016\\-.0018\\\vdots\\.0038\end{bmatrix}$	10.9087	24.5064	0.4451	0.4451	$\begin{bmatrix}-0.0032\\0.4451\\0\\0\end{bmatrix}$
$i = 3$	$\begin{bmatrix}-0.0032\\0.4451\\0\\0\end{bmatrix}$	$\begin{bmatrix}-1.100\\0.4574\\\vdots\\-0.8085\end{bmatrix}$	$\begin{bmatrix}-0.0369\\0.0524\\\vdots\\-0.6280\end{bmatrix}$	15.9404	23.4301	0.6803	0.6803	$\begin{bmatrix}-0.0032\\0.4451\\0.6803\\0\end{bmatrix}$
$i = 4$	$\begin{bmatrix}-0.0032\\0.4451\\0.6803\\0\end{bmatrix}$	$\begin{bmatrix}1.0441\\-0.2866\\\vdots\\-0.8524\end{bmatrix}$	$\begin{bmatrix}-0.7853\\0.3636\\\vdots\\-1.1781\end{bmatrix}$	1.2659	24.0035	0.0527	0.0527	$\begin{bmatrix}-0.0032\\0.4451\\0.6803\\0.0527\end{bmatrix}$

TABLE 5.13
Estimated Parameters for Different Values of λ

λ	$\hat{\beta}_1$	$\hat{\beta}_2$	$\hat{\beta}_3$	$\hat{\beta}_4$
0.0020	0.4943	−0.4920	−0.1361	1.2909
0.0424	0.3724	−0.3565	−0.0966	1.1831
0.0828	0.2504	−0.2210	−0.0571	1.0752
0.1231	0.1285	−0.0854	−0.0175	0.9674
0.1635	0.0515	0	0.0	0.9062
0.2039	0.0507	0	0	0.9053
0.2443	0.0498	0	0	0.9044
0.2847	0.0489	0	0	0.9035
0.3250	0.0480	0	0	0.9026
0.3654	0.0471	0	0	0.9017
0.4058	0.0462	0	0	0.9008
0.4462	0.0454	0	0	0.8999
0.4866	0.0445	0	0	0.8990
0.5269	0.0436	0	0	0.8981
0.5673	0.0427	0	0	0.8972
0.6077	0.0418	0	0	0.8963
0.6481	0.0409	0	0	0.8954
0.6885	0.0400	0	0	0.8945
0.7288	0.0392	0	0	0.8936
0.7692	0.0383	0	0	0.8927
0.8096	0.0374	0	0	0.8918
0.8500	0.0365	0	0	0.8909
0.8904	0.0356	0	0	0.8900
0.9307	0.0208	0.0156	0	0.8811
0.9711	0.0020	0.0358	0	0.8699
1.0115	0	0.0374	0	0.8683
1.0519	0	0.0368	0	0.8677
1.0923	0	0.0362	0	0.8671
1.1326	0	0.0357	0	0.8665
1.1730	0	0.0351	0	0.8659
1.2134	0	0.0345	0	0.8653
1.2538	0	0.0340	0	0.8647
1.2942	0	0.0334	0	0.8642
1.3345	0	0.0328	0	0.8636
1.3749	0	0.0323	0	0.8630
1.4153	0	0.0317	0	0.8624
1.4557	0	0.0311	0	0.8618
1.4961	0	0.0306	0	0.8612
1.5364	0	0.0300	0	0.8606
1.5768	0	0.0295	0	0.8600
1.6172	0	0.0289	0	0.8595
1.6576	0	0.0283	0	0.8589

(Continued)

TABLE 5.13 (*Continued*)
Estimated Parameters for Different Values of λ

λ	$\hat{\beta}_1$	$\hat{\beta}_2$	$\hat{\beta}_3$	$\hat{\beta}_4$
1.6980	0	0.0278	0	0.8583
1.7383	0	0.0272	0	0.8577
1.7787	0	0.0266	0	0.8571
1.8191	0	0.0261	0	0.8565
1.8595	0	0.0255	0	0.8559
1.8999	0	0.0249	0	0.8554
1.9402	0	0.0244	0	0.8548
1.9806	0	0.0238	0	0.8542
2.0210	0	0.0232	0	0.8536
2.0614	0	0.0227	0	0.8530
2.1018	0	0.0221	0	0.8524
2.1421	0	0.0215	0	0.8518
2.1825	0	0.0210	0	0.8513
2.2229	0	0.0204	0	0.8507
2.2633	0	0.0199	0	0.8501
2.3037	0	0.0193	0	0.8495
2.3440	0	0.0187	0	0.8489
2.3844	0	0.0182	0	0.8483
2.4248	0	0.0176	0	0.8477
2.4652	0	0.0170	0	0.8472
2.5056	0	0.0165	0	0.8466
2.5459	0	0.0159	0	0.8460
2.5863	0	0.0153	0	0.8454
2.6267	0	0.0148	0	0.8448
2.6671	0	0.0142	0	0.8442
2.7075	0	0.0136	0	0.8436
2.7478	0	0.0131	0	0.8431
2.7882	0	0.0125	0	0.8425
2.8260	0	0.0120	0	0.8419
2.8690	0	0.0114	0	0.8413
2.9094	0	0.0108	0	0.8407
2.9497	0	0.0103	0	0.8401
2.9901	0	0.0097	0	0.8395
3.0305	0	0.0091	0	0.8390
3.0709	0	0.0086	0	0.8384
3.1113	0	0.0080	0	0.8378
3.1516	0	0.0074	0	0.8372
3.1920	0	0.0069	0	0.8366
3.2324	0	0.0063	0	0.8360
3.2728	0	0.0057	0	0.8354
3.3132	0	0.0052	0	0.8349
3.3535	0	0.0046	0	0.8343

(*Continued*)

TABLE 5.13 (*Continued*)
Estimated Parameters for Different Values of λ

λ	$\hat{\beta}_1$	$\hat{\beta}_2$	$\hat{\beta}_3$	$\hat{\beta}_4$
3.3939	0	0.0041	0	0.8337
3.4343	0	0.0035	0	0.8331
3.4747	0	0.0029	0	0.8325
3.5151	0	0.0024	0	0.8319
3.5554	0	0.0018	0	0.8313
3.5958	0	0.0012	0	0.8307
3.6362	0	0.0007	0	0.8302
3.6766	0	0	0	0.8296
3.7170	0	0	0	0.8288
3.7573	0	0	0	0.8279
3.7977	0	0	0	0.8271
3.8381	0	0	0	0.8263
3.8785	0	0	0	0.8254
3.9189	0	0	0	0.8246
3.9592	0	0	0	0.8237
4.000	0	0	0	0.8229

7. Calculate the validation error of these selected models for the validation set as shown in Table 5.14. For example,

$$\hat{\mathbf{y}} = \begin{bmatrix} 0.3902 & -0.0662 & -1.0149 & 0.0013 \\ -0.5366 & -0.44 & 0.5362 & 0.4417 \\ -0.3101 & -0.5518 & -0.4277 & -0.4676 \end{bmatrix} \times \begin{bmatrix} 0.0515 \\ 0 \\ 0 \\ 0.9062 \end{bmatrix}$$

$$= \begin{bmatrix} -0.6902 \\ 0.3726 \\ -0.4397 \end{bmatrix}$$

$$\text{Validation error} = \sum_{i \in validationset} (y_i - \hat{y}_i)^2 = 0.5821$$

8. Select the model with the minimum value of validation error. Hence, the final estimated answer would be

$$\hat{\beta} = [0, 0.007, 0, 0.8302]$$

Note that without deciding how many variables will be used, one can also choose the one with the smallest validation error. For example, the validation errors are calculated for all the λ values (Table 5.15). Comparing the validation errors, listed in Table 5.15, $\lambda = 4$ and its related estimation can be selected as the final result. This selected model has the minimum validation error among all the studied models.

TABLE 5.14
Validation Errors for Vectors of Parameters with Two Nonzero Elements

$\hat{\beta}_1$	$\hat{\beta}_2$	$\hat{\beta}_3$	$\hat{\beta}_4$	Validation Error
0.0515	0	0	0.9062	0.5821
0.0507	0	0	0.9053	0.5818
0.0498	0	0	0.9044	0.5817
0.0489	0	0	0.9035	0.5815
0.0480	0	0	0.9026	0.5814
0.0471	0	0	0.9017	0.5812
0.0462	0	0	0.9008	0.5810
0.0454	0	0	0.8999	0.5809
0.0445	0	0	0.8990	0.5807
0.0436	0	0	0.8981	0.5806
0.0427	0	0	0.8972	0.5804
0.0418	0	0	0.8963	0.5803
0.0409	0	0	0.8954	0.5801
0.0400	0	0	0.8945	0.5800
0.0392	0	0	0.8936	0.5798
0.0383	0	0	0.8927	0.5797
0.0374	0	0	0.8918	0.5795
0.0365	0	0	0.8909	0.5794
0.0356	0	0	0.8900	0.5792
0	0.0374	0	0.8683	0.5752
0	0.0368	0	0.8677	0.5748
0	0.0362	0	0.8671	0.5744
0	0.0357	0	0.8665	0.5740
0	0.0351	0	0.8659	0.5735
0	0.0345	0	0.8653	0.5731
0	0.0340	0	0.8647	0.5727
0	0.0334	0	0.8642	0.5723
0	0.0328	0	0.8636	0.5719
0	0.0323	0	0.8630	0.5715
0	0.0317	0	0.8624	0.5711
0	0.0311	0	0.8618	0.5707
0	0.0306	0	0.8612	0.5702
0	0.0300	0	0.8606	0.5698
0	0.0295	0	0.8600	0.5694
0	0.0289	0	0.8595	0.5690
0	0.0283	0	0.8589	0.5686
0	0.0278	0	0.8583	0.5682
0	0.0272	0	0.8577	0.5678
0	0.0266	0	0.8571	0.5674
0	0.0261	0	0.8565	0.5670
0	0.0255	0	0.8559	0.5666

(Continued)

TABLE 5.14 (*Continued*)
Validation Errors for Vectors of Parameters with Two Nonzero Elements

$\hat{\beta}_1$	$\hat{\beta}_2$	$\hat{\beta}_3$	$\hat{\beta}_4$	Validation Error
0	0.0249	0	0.8554	0.5662
0	0.0244	0	0.8548	0.5658
0	0.0238	0	0.8542	0.5654
0	0.0232	0	0.8536	0.5650
0	0.0227	0	0.8530	0.5646
0	0.0221	0	0.8524	0.5643
0	0.0215	0	0.8518	0.5639
0	0.0210	0	0.8513	0.5635
0	0.0204	0	0.8507	0.5631
0	0.0199	0	0.8501	0.5627
0	0.0193	0	0.8495	0.5623
0	0.0187	0	0.8489	0.5619
0	0.0182	0	0.8483	0.5615
0	0.0176	0	0.8477	0.5612
0	0.0170	0	0.8472	0.5608
0	0.0165	0	0.8466	0.5604
0	0.0159	0	0.8460	0.5600
0	0.0153	0	0.8454	0.5596
0	0.0148	0	0.8448	0.5593
0	0.0142	0	0.8442	0.5589
0	0.0136	0	0.8436	0.5585
0	0.0131	0	0.8431	0.5581
0	0.0125	0	0.8425	0.5578
0	0.0120	0	0.8419	0.5574
0	0.0114	0	0.8413	0.5570
0	0.0108	0	0.8407	0.5567
0	0.0103	0	0.8401	0.5563
0	0.0097	0	0.8395	0.5559
0	0.0091	0	0.8390	0.5556
0	0.0086	0	0.8384	0.5552
0	0.0080	0	0.8378	0.5548
0	0.0074	0	0.8372	0.5545
0	0.0069	0	0.8366	0.5541
0	0.0063	0	0.8360	0.5538
0	0.0057	0	0.8354	0.5534
0	0.0052	0	0.8349	0.5531
0	0.0046	0	0.8343	0.5527
0	0.0041	0	0.8337	0.5523
0	0.0035	0	0.8331	0.5520
0	0.0029	0	0.8325	0.5516
0	0.0024	0	0.8319	0.5513
0	0.0018	0	0.8313	0.5509
0	0.0012	0	0.8307	0.5506
0	0.0007	0	0.8302	0.5503

TABLE 5.15
Values of Validation Error for Different Values of λ

λ	Validation Error	λ	Validation Error	λ	Validation Error
0.0020	0.5740	1.3749	0.5715	2.7478	0.5581
0.0424	0.5756	1.4153	0.5711	2.7882	0.5578
0.0828	0.5784	1.4557	0.5707	2.8260	0.5574
0.1231	0.5824	1.4961	0.5702	2.8690	0.5570
0.1635	0.5821	1.5364	0.5698	2.9094	0.5567
0.2039	0.5818	1.5768	0.5694	2.9497	0.5563
0.2443	0.5817	1.6172	0.5690	2.9901	0.5559
0.2847	0.5815	1.6576	0.5686	3.0305	0.5556
0.3250	0.5814	1.6980	0.5682	3.0709	0.5552
0.3654	0.5812	1.7383	0.5678	3.1113	0.5548
0.4058	0.5810	1.7787	0.5674	3.1516	0.5545
0.4462	0.5809	1.8191	0.5670	3.1920	0.5541
0.4866	0.5807	1.8595	0.5666	3.2324	0.5538
0.5269	0.5806	1.8999	0.5662	3.2728	0.5534
0.5673	0.5804	1.9402	0.5658	3.3132	0.5531
0.6077	0.5803	1.9806	0.5654	3.3535	0.5527
0.6481	0.5801	2.0210	0.5650	3.3939	0.5523
0.6885	0.5800	2.0614	0.5646	3.4343	0.5520
0.7288	0.5798	2.1018	0.5643	3.4747	0.5516
0.7692	0.5797	2.1421	0.5639	3.5151	0.5513
0.8096	0.5795	2.1825	0.5635	3.5554	0.5509
0.8500	0.5794	2.2229	0.5631	3.5958	0.5506
0.8904	0.5792	2.2633	0.5627	3.6362	0.5503
0.9307	0.5777	2.3037	0.5623	3.6766	0.5499
0.9711	0.5758	2.3440	0.5619	3.7170	0.5492
1.0115	0.5752	2.3844	0.5615	3.7573	0.5483
1.0519	0.5748	2.4248	0.5612	3.7977	0.5475
1.0923	0.5744	2.4652	0.5608	3.8381	0.5467
1.1326	0.5740	2.5056	0.5604	3.8785	0.5459
1.1730	0.5735	2.5459	0.5600	3.9189	0.5450
1.2134	0.5731	2.5863	0.5596	3.9592	0.5442
1.2538	0.5727	2.6267	0.5593	4.000	0.5434
1.2942	0.5723	2.6671	0.5589		
1.3345	0.5719	2.7075	0.5585		

5.3 NONLINEAR REGRESSION MODELING

5.3.1 Artificial Neural Network

An artificial neural network (ANN) is a system of interconnected computational elements mimicking the biological nerve system consisting of input, hidden, and output layers, which include nodes (or neurons). The ANN has been popularly employed in meteorological and hydrological sciences for forecasting and simulating the related variables. The nonlinear regression with the ANN technique can be easily adopted to regression-based statistical downscaling as a simple linear regression.

The ANN contains key elements, such as an input layer, a hidden layer, and an output layer. At each layer, there are a certain number of neurons (or nodes). A simple diagram of the simple ANN architecture is presented in Figure 5.8 with two nodes in the input layer, three nodes in the hidden layer, and one node in the output layer. When the input nodes receive data in the input layer, they pass to the hidden layer and finally to the output layer.

While passing the data, the ANN acts a nonlinear (or sometimes linear) function with a certain weight. The function is called "an activation function" with various forms such as hyperbolic tangents, sigmoids, and polynomials. A sigmoid (σ) and hyperbolic tangent (tanh) can be defined as

$$\sigma(t) = (1 + e^{-t})^{-1} \tag{5.29}$$

$$\tanh(t) = \frac{e^t - e^{-t}}{e^t + e^{-t}} \tag{5.30}$$

If a sigmoid function at the input and hidden layers passage and a linear function at the hidden output layer are applied for the example of Figure 5.8, the ANN architecture can be explicitly described as a function:

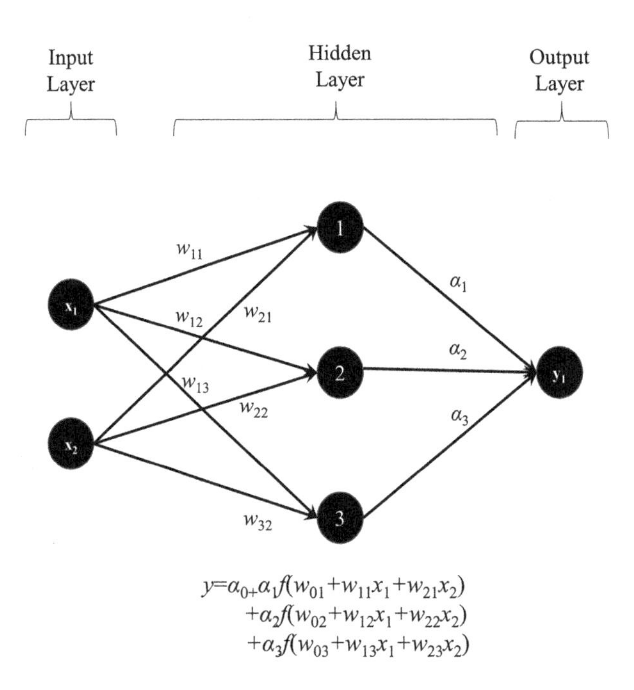

FIGURE 5.8 A simple example of the ANN architecture with two inputs and three hidden neurons on the output layer.

$$y_i = \alpha_0 + \alpha_1\left(\left[1+\exp\{-(w_{01}+x_{1,i}w_{11}+x_{2,i}w_{21})\}\right]^{-1}\right)$$
$$+\alpha_2\left(\left[1+\exp\{-(w_{02}+x_{1,i}w_{12}+x_{2,i}w_{22})\}\right]^{-1}\right) \quad (5.31)$$
$$+\alpha_3\left(\left[1+\exp\{-(w_{03}+x_{1,i}w_{13}+x_{2,i}w_{23})\}\right]^{-1}\right)$$

Note that w_{uv} (u = 1, 2; v = 1, 2, 3) and α_v (v = 1, 2 ,3) are the weights (or parameters) that provide the importance of connections between nodes with offsets (or bias, i.e., w_{0v} and α_0).

Since the network requires a complex architecture setup and parameter estimation, the ANN employs the training and validation algorithm, with the datasets separated into three parts (i.e., training, validation, and testing periods). The overall procedure with the three parts of the data is as follows:

1. Estimate the parameters for a given ANN architecture with the training dataset.
2. Check the performance of the architecture with the estimated parameters (or weights) and the validation dataset. The architecture can be manipulated at this stage to meet some preestablished criteria.
3. Test the final architecture and the estimated parameters with the test dataset.

The parameters are estimated by minimizing the approximation square error (ASE) defined as

$$E = \sum_{i=1}^{T} E_i = \sum_{i=1}^{T} \frac{1}{2}(\hat{y}_i - y_i)^2 \quad (5.32)$$

where T is the number of training data, y_i is the observation, and $\hat{y}_i$ is the estimate from the ANN model as in Eq. (5.31). The minimum of E is found by equating the partial derivatives of E to zero, i.e.,

$$\frac{\partial E}{\partial \theta_p} = 0, \quad p = 1,\ldots P \quad (5.33)$$

where P is the number of parameters.

In the ANN literature, the back-propagation scheme has been popularly employed by adjusting the parameters along the negative gradient of the ASE. By initializing the parameter set $\boldsymbol{\theta}$ as $\boldsymbol{\theta}^{(0)}$, the next value of $\boldsymbol{\theta}$ is updated as

$$\boldsymbol{\theta}^{(k+1)} = \boldsymbol{\theta}^{(k)} - \zeta \left.\frac{\partial E}{\partial \boldsymbol{\theta}}\right|_{\boldsymbol{\theta}=\boldsymbol{\theta}^{(k)}} \quad (5.34)$$

where ζ is the learning rate.

The updating of parameters can be done either sequentially as online training or over the entire parameter set as batch-mode training. A number of parameter estimation algorithms have been developed, such as conjugate gradient algorithm (Wasserman, 1993) and Bayesian regularization (Mackay, 1992).

Furthermore, the parameters (or weights) can be estimated with metaheuristic algorithms, such as Genetic Algorithm (Srinivasulu and Jain, 2006; Nasseri et al., 2008), Particle Swarm Optimization (Sedki and Ouazar, 2010), or Harmony Search (Lee et al., 2016). Mostly, the metaheuristic algorithms search the parameter set that results in the optimum of the objective function [here, minimum of Eq. (5.32)] by generating a number of possible candidates.

To check the performance of the given architecture and the estimated parameter set with the validation data ($i = 1, \ldots, V$; V is the number of validation data), some criteria must be established such as bias and squared coefficient of correlation (R^2):

$$Bias = \frac{1}{V}\sum_{i=1}^{V}(\hat{y}_i - y_i) = \mu_{\hat{y}} - \mu_y \tag{5.35}$$

The coefficient of determination R^2 is

$$R^2 = \left[\frac{\sum_{i=1}^{V}(\hat{y}_i - \mu_{\hat{y}})(y_i - \mu_y)}{\sqrt{\sum_{i=1}^{V}(\hat{y}_i - \mu_{\hat{y}})^2}\sqrt{\sum_{i=1}^{V}(y_i - \mu_y)^2}}\right]^2 \tag{5.36}$$

Example 5.6

Downscale the January maximum precipitation at Cedars, Quebec, with part of the dataset in Table 5.4 as specific humidity and near-surface air temperature (see the fourth and fifth columns) using ANN.

Solution:

The same standardization is required in Table 5.5 for the same reason as in the multiple linear regression. Also, parameter estimation is much simpler with the standardized data, since parameters are restricted within a certain range.

1. Preprocess the standardized dataset in Table 5.5 (specific humidity and near-surface air temperature for input layer and the January maximum temperature for output layer) by separating the dataset as training (1961–1981, 20 years) and validation periods (1981–1990, 6 years). Note that no testing period is assigned in the current example, since its record length is too short.
2. Create an architecture that is the same as in Figure 5.8. Three nodes are assigned to the hidden layer and a sigmoid function to the input and hidden layer passage, and a linear function to the hidden and output passage are applied. Also, the architectures with one and two hidden nodes are tested in this example.

3. Estimate the parameters [w_{uv} (u = 0, 1, 2; v = 1, 2, 3) and α_k (k = 0, 1, 2, 3)] with an appropriate estimation approach. Here, the metaheuristic algorithm, Harmony Search, is applied with the objective function in Eq. (5.32). The detailed parameter estimation procedure is omitted, since its procedure and example are explained in Chapter 2. The estimated parameters are

$$\alpha = [-0.81, -0.62, 0.93, 1.17]$$

$$\mathbf{w} = \begin{bmatrix} 2.17 & -4.07 & 6.87 \\ 3.68 & 1.41 & -7.06 \\ -1.09 & 6.17 & -8.80 \end{bmatrix}$$

4. Calculate the predictand estimate ($\hat{y}_i$) for the training period by applying these parameters. For example, the estimate of the first year (1961, i = 1) with $x_{11} = -1.101$ and $x_{12} = -1.042$ is

$$\hat{y}_1 = -0.81 - 0.62\left(\left[1 + \exp\{-(2.17 + (-1.101)\times(-4.07) + (-1.042)\times(6.87))\}\right]^{-1}\right)$$
$$+ 0.93\left(\left[1 + \exp\{-(3.68 + (-1.101)\times 1.41 + (-1.042)\times(-7.06))\}\right]^{-1}\right)$$
$$+ 1.17\left(\left[1 + \exp\{-(-1.09 + (-1.101)\times 6.17 + (-1.042)\times(-8.8))\}\right]^{-1}\right)$$
$$= -0.9396$$

All the estimates for the training and validation periods are shown for the three hidden node case in the second and sixth columns in Table 5.16.

5. Estimate the performance criteria [i.e., Bias in Eq. (5.35) and R^2 in Eq. (5.36)] as well as the objective function of ASE (E) in Eq. (5.32) for the three hidden nodes. The estimates are as –0.003, 0.939, and 0.55, respectively.
6. Vary the architecture to find the best performance by increasing or decreasing the number of hidden nodes and changing the activation function. Here, different numbers of hidden nodes (one and two hidden nodes) are assigned to the same architecture as the three hidden node case. Follow steps (3)–(5)
7. Compare the performance criteria for all the different architectures. Table 5.16 and Figure 5.9 reveal that the architecture with two hidden nodes present the best performance for E and R^2 with slightly larger bias than the one for the architecture of the one hidden node. Therefore, the architecture with two hidden nodes can be selected.

   Note that the main reason to separate the dataset into three parts as the training, validation, and test period is to estimate the parameters with the training data, update the architecture with the validation data, and check the final performance with the test period. The data separation requires a significant record length, and the current example has the limited dataset so that the test period is omitted.

TABLE 5.16
Nonliear Regression Estimate for Training and Validation Periods with the ANN of the Same Architecture as in Figure 5.8 as well as Two Hidden Nodes and One Hidden Node for the Target January Maximum Temperature (°C) at Cedars, Quebec, with the Two Atmospheric Input Variables of Specific Humidity and Near-Surface Air Temperature (see Tables 5.4 and 5.5)

Train	3 Hid. Nodes	2 Hid. Nodes	1 Hid. Nodes	Valid	3 Hid. Nodes	2 Hid. Nodes	1 Hid. Nodes
1961	–0.940	–0.879	–1.016	1984	0.118	–0.003	0.120
1962	0.075	–0.034	0.050	1985	0.085	0.009	–0.080
1963	0.151	0.042	0.082	1986	0.010	0.183	0.185
1964	1.128	1.210	0.959	1987	1.034	0.976	0.995
1965	–0.256	–0.226	–0.507	1989	1.212	1.226	1.018
1967	1.273	1.223	1.041	1990	1.292	1.234	1.099
1968	–1.290	–1.300	–1.241	E	1.370	1.154	1.287
1969	0.638	0.383	0.790	R^2	0.676	0.814	0.700
1970	–1.335	–1.302	–1.316	Bias	0.259	0.238	0.190
1971	–0.942	–0.940	–0.971				
1972	0.650	0.879	0.698				
1973	0.765	0.854	0.580				
1974	0.542	0.543	0.655				
1975	1.241	1.227	1.023				
1976	–1.167	–1.193	–1.128				
1977	–1.312	–1.295	–1.283				
1978	–0.528	–0.580	–0.569				
1979	0.352	0.295	0.490				
1980	0.114	–0.045	0.212				
1981	–1.416	–1.383	–1.405				
E	0.550	0.510	0.742				
R^2	0.939	0.953	0.921				
Bias	–0.003	–0.016	–0.033				

Note that the smaller E and Bias the better performance while the larger R^2 the better.

5.4 SUMMARY AND CONCLUSION

Functional relationships between GCM climate variables and local weather-station variables are employed to downscale the GCM output scenarios to local weather variables. Multiple linear regression models have commonly been used, accompanied with the predictor selection methods, such as SWR and LASSO. Nonlinear relations are difficult to capture. Recent machine-learning techniques, such as ANN and support vector machine, have allowed easy access to model nonlinearity. In this chapter, ANN is introduced and presented with an example to show its implementation.

Note that complex models such as nonlinear models and a large number of predictor variables might have a better representation for downscaling. However, it

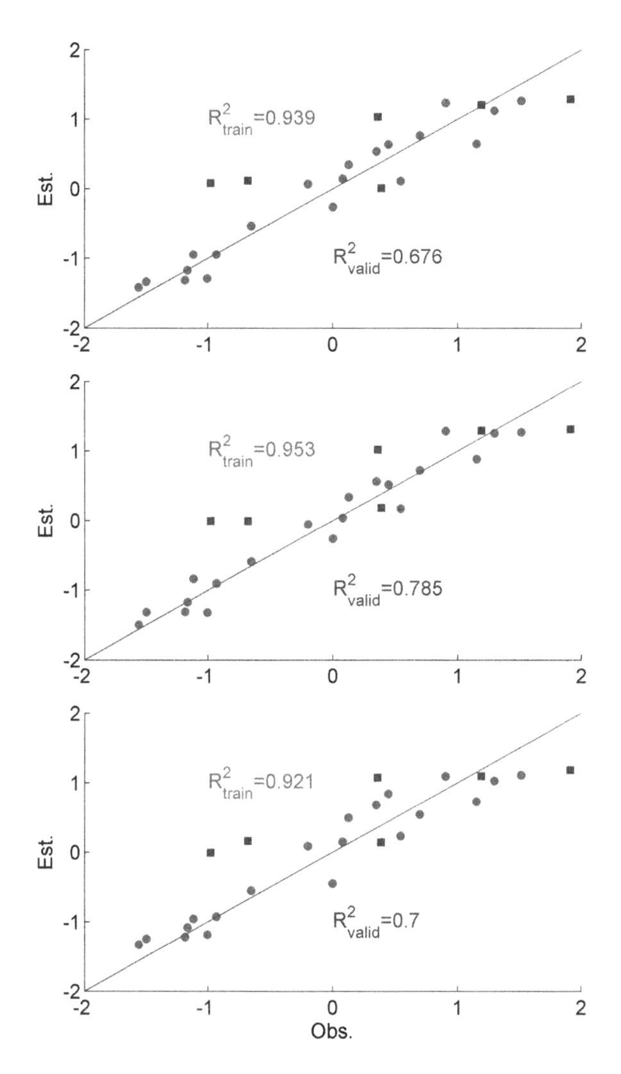

FIGURE 5.9 Performance of the neural network for the same architecture as in Figure 5.8 (top panel) as well as two hidden nodes (middle panel) and one hidden node (bottom panel) for the target January maximum temperature (°C) at Cedars, Quebec, and the two atmospheric predictors of specific humidity and near-surface air temperature (see Tables 5.4 and 5.5).

is easy to overfit for downscaling and might not work well enough in a real application for future precipitation, and caution must be taken in such a case. Furthermore, a clear linear or nonlinear relationship tends not to show in the precipitation variable, since its natural rain-generation algorithm is too complex to model in GCMs and also the ground measurement might include some uncertainties and an employed weather station might have local influences such as orographic effect and not be representative. In contrast, the temperature variable often presents high association between climate variables, which are outputs of GCM, and local measurements. This model has been more likely employed for a temperature variable.

6 Weather Generator Downscaling

6.1 MATHEMATICAL BACKGROUND

6.1.1 AUTOREGRESSIVE MODELS

The autoregressive model of order p (AR(p)) for a time-series variable Y_t can be generally described as

$$Y_t = \sum_{j=1}^{p} \varphi_j Y_{t-j} + \varepsilon_t \tag{6.1}$$

where Y_t is the time-dependent series with zero mean and ε_t is the time-independent series as $\varepsilon_t \sim N\left(0, \sigma_\varepsilon^2\right)$. The simple AR(1) model, whose order (p) is one for AR(p), can be expressed as

$$Y_t = \varphi_1 Y_{t-1} + \varepsilon_t \tag{6.2}$$

To apply a time-series model, the target time series must follow the assumption such that there is no long-term trend (such as consistent increasing or decreasing trends) or cycles (periodic changes of daily or annual). Furthermore, the time-series variable Y_t is assumed to be normally distributed. To meet this normality, the target time series must be transformed by log or power function.

6.1.2 MULTIVARIATE AUTOREGRESSIVE MODEL

The simple multivariate autoregressive model, MAR(1), with S number of variables can be described as

$$\mathbf{Y}_t = \mathbf{A}_1 \mathbf{Y}_{t-1} + \mathbf{B}\mathbf{E}_t \tag{6.3}$$

where $\mathbf{Y}_t$ and $\mathbf{E}_t$ are vectors representing $\left[Y_t^1, Y_t^2, \ldots Y_t^S\right]^T$ and $\left[\varepsilon_t^1, \varepsilon_t^2, \ldots \varepsilon_t^S\right]^T$, respectively, and $\mathbf{A}_1$ and $\mathbf{B}$ are the parameter matrices ($S \times S$). This equation may be expanded as

$$\begin{bmatrix} Y_t^1 \\ Y_t^2 \\ \vdots \\ Y_t^S \end{bmatrix} = \begin{bmatrix} a_{11} & a_{12} & \cdots & a_{1S} \\ a_{21} & a_{22} & \cdots & a_{2S} \\ \vdots & \vdots & \ddots & \vdots \\ a_{S1} & a_{S2} & \cdots & a_{SS} \end{bmatrix} \begin{bmatrix} Y_{t-1}^1 \\ Y_{t-1}^2 \\ \vdots \\ Y_{t-1}^S \end{bmatrix} + \begin{bmatrix} b_{11} & b_{12} & \cdots & b_{1S} \\ b_{21} & b_{22} & \cdots & b_{2S} \\ \vdots & \vdots & \ddots & \vdots \\ b_{S1} & b_{S2} & \cdots & b_{SS} \end{bmatrix} \begin{bmatrix} \varepsilon_t^1 \\ \varepsilon_t^2 \\ \vdots \\ \varepsilon_t^S \end{bmatrix} \tag{6.4}$$

The moment estimates of parameter matrices by multiplying $\mathbf{Y}_t$ and $\mathbf{Y}_{t-1}$ and taking expectations may be obtained as (Matalas, 1967):

$$E\left[\mathbf{Y}_t\mathbf{Y}_t^T\right] = \mathbf{A}_1 E\left[\mathbf{Y}_{t-1}\mathbf{Y}_t^T\right] + \mathbf{B}E\left[\mathbf{E}_t\mathbf{Y}_t^T\right] \tag{6.5}$$

$$\mathbf{M}_0 = \mathbf{A}_1\mathbf{M}_1 + \mathbf{B}\mathbf{B}^T \tag{6.6}$$

and

$$E\left[\mathbf{Y}_t\mathbf{Y}_{t-1}^T\right] = \mathbf{A}_1 E\left[\mathbf{Y}_{t-1}\mathbf{Y}_{t-1}^T\right] + \mathbf{B}E\left[\mathbf{E}_t\mathbf{Y}_{t-1}^T\right] \tag{6.7}$$

$$\mathbf{M}_1 = \mathbf{A}_1\mathbf{M}_0 \tag{6.8}$$

where $\mathbf{M}_0 = E\left[\mathbf{Y}_t\mathbf{Y}_t^T\right]$ and $\mathbf{M}_1 = E\left[\mathbf{Y}_t\mathbf{Y}_{t-1}^T\right]$. Note that $\mathbf{E}_t$ are time independent and $E\left[\mathbf{E}_t\mathbf{Y}_{t-1}^T\right] = \mathbf{0}$. From these two equations,

$$\mathbf{A}_1 = \mathbf{M}_1\mathbf{M}_0^{-1} \tag{6.9}$$

and

$$\mathbf{B}\mathbf{B}^T = \mathbf{M}_0 - \mathbf{A}_1\mathbf{M}_1^{\,T} \tag{6.10}$$

where $\mathbf{M}_0$ and $\mathbf{M}_1$ are, respectively, the lag-zero and lag-one correlation matrices of $\mathbf{Y}_t$:

$$\mathbf{M}_0 = \begin{bmatrix} corr(y_t^1, y_t^1) & corr(y_t^1, y_t^2) & \cdots & corr(y_t^1, y_t^S) \\ corr(y_t^2, y_t^1) & corr(y_t^2, y_t^2) & \cdots & corr(y_t^2, y_t^S) \\ \vdots & \vdots & \ddots & \vdots \\ corr(y_t^S, y_t^1) & corr(y_t^S, y_t^2) & \cdots & corr(y_t^S, y_t^S) \end{bmatrix}$$

and

$$\mathbf{M}_1 = \begin{bmatrix} corr(y_t^1, y_{t-1}^1) & corr(y_t^1, y_{t-1}^2) & \cdots & corr(y_t^1, y_{t-1}^S) \\ corr(y_t^2, y_{t-1}^1) & corr(y_t^2, y_{t-1}^2) & \cdots & corr(y_t^2, y_{t-1}^S) \\ \vdots & \vdots & \ddots & \vdots \\ corr(y_t^S, y_{t-1}^1) & corr(y_t^S, y_{t-1}^2) & \cdots & corr(y_t^S, y_{t-1}^S) \end{bmatrix}$$

See Eq. (2.57) for correlation. These matrices can simply be estimated with the observed dataset of $\vec{\mathbf{y}}$ and $\tilde{\mathbf{y}}$ as

$$\hat{\mathbf{M}}_0 = \frac{1}{n-1}\vec{\mathbf{y}}\vec{\mathbf{y}}^T \text{ and } \hat{\mathbf{M}}_1 = \frac{1}{n-1}\vec{\mathbf{y}}\tilde{\mathbf{y}}^T \tag{6.11}$$

where

$$\vec{\mathbf{y}} = \begin{bmatrix} y_2^1 & y_3^1 & \cdots & y_n^1 \\ y_2^2 & y_3^2 & \cdots & y_n^2 \\ \vdots & \vdots & \ddots & \vdots \\ y_2^S & y_3^S & \cdots & y_n^S \end{bmatrix} \text{ and } \tilde{\mathbf{y}} = \begin{bmatrix} y_1^1 & y_2^1 & \cdots & y_{n-1}^1 \\ y_1^2 & y_2^2 & \cdots & y_{n-1}^2 \\ \vdots & \vdots & \ddots & \vdots \\ y_1^S & y_2^S & \cdots & y_{n-1}^S \end{bmatrix}$$

Note that the record length of $n - 1$ is used to take lag-one into account instead of n.

If $\mathbf{B}$ is a lower triangular matrix but $\mathbf{D} = \mathbf{BB}^T$ [see the left side of (6.10)] is a positive definite matrix, then the elements of $\mathbf{B}$ may be determined as

$$b_{ij} = 0 \quad \text{for all } i < j \tag{6.12}$$

$$b_{11} = \sqrt{d_{11}} \text{ and } b_{i1} = \frac{d_{1i}}{\sqrt{d_{11}}} \quad \text{for } i = 1,\ldots,S \tag{6.13}$$

$$b_{ii} = \left[d_{ii} - \left(b_{i1}^2 + b_{i2}^2 + \cdots + b_{ii-1}^2\right)\right]^{1/2} \quad \text{for } i = 2,\ldots,S \tag{6.14}$$

$$b_{ji} = \frac{d_{ij} - \left(b_{i1}b_{j1} + b_{i2}b_{j2} + \cdots + b_{ii-1}b_{ji-1}\right)}{b_{ii}} \quad \text{for } j > i \tag{6.15}$$

Overall, the parameter matrix $\mathbf{B}$ is

$$\mathbf{B} = \begin{bmatrix} b_{11} & b_{12} & \cdots & b_{1S} \\ b_{21} & b_{22} & \cdots & b_{2S} \\ \vdots & \vdots & \ddots & \vdots \\ b_{S1} & b_{S2} & \cdots & b_{SS} \end{bmatrix}$$

$$= \begin{bmatrix} \sqrt{d_{11}} & 0 & \cdots & 0 \\ \frac{d_{12}}{\sqrt{d_{11}}} & \left[d_{22} - \left(b_{21}^2\right)\right]^{1/2} & \cdots & 0 \\ \vdots & \vdots & \ddots & \vdots \\ \frac{d_{1S}}{\sqrt{d_{11}}} & \frac{d_{2S} - \left(b_{S1}b_{21} + b_{S2}b_{22} + \cdots + b_{SS-1}b_{S2-1}\right)}{b_{22}} & \cdots & \left[d_{SS} - \left(b_{S1}^2 + b_{S2}^2 + \cdots + b_{SS-1}^2\right)\right]^{1/2} \end{bmatrix} \tag{6.16}$$

6.1.3 Markov Chain

A Markov chain is a sequence of random variables, such that the probability of moving to the next state depends only on the present state and not on the previous states. The precipitation occurrence has two states as wet or dry, and the probability of a wet day given the past states can be described with a Markov chain model as

$$\Pr\left(X_t = 1 \mid X_{t-1} = a_{t-1}, \dots, X_1 = a_1\right) = \Pr\left(X_t = 1 \mid X_{t-1} = a_{t-1}\right) = P_{a_{t-1},1} \quad (6.17)$$

where $a_{t-1}, \dots, a_1$ are the occurrence states of the past conditions either 0 or 1. The transition probability matrix (P_T) is

$$\mathbf{P}_T = \begin{bmatrix} P_{00} & P_{01} \\ P_{10} & P_{11} \end{bmatrix} \quad (6.18)$$

Note that $P_{00} = 1 - P_{01}$ and $P_{11} = 1 - P_{10}$. These probabilities are estimated by counting the number of events that follow this condition. For example,

$$P_{01} = \frac{N\left(x_t = 1 \mid x_{t-1} = 0\right)}{N\left(x_{t-1} = 0\right)} = \frac{N_{01}}{N_1} \quad (6.19)$$

where $N(\Delta)$ is the number of events meeting the condition Δ.

Also, the occurrence probability without the previous condition (P_0 or P_1), sometimes called as limiting probabilities ($\mathbf{P}_L = [P_0, P_1]^T$), can be estimated from a simple algebra with $\mathbf{P}_L = \mathbf{P}_T \times \mathbf{P}_L$ as

$$P_1 = \frac{P_{01}}{1 - P_{11} + P_{01}} = \frac{P_{01}}{P_{10} + P_{01}} \quad \text{and} \quad P_0 = \frac{1 - P_{11}}{1 - P_{11} + P_{01}} = \frac{P_{10}}{P_{10} + P_{01}} \quad (6.20)$$

Example 6.1

Estimate the probability transition matrix in Eq. (6.18) for the daily precipitation dataset given in Table 6.1 for the month of July 2002 from Seoul.

Solution:

In the second column of Table 6.1, its occurrence is presented as a Bernoulli process (i.e., wet = 1 or dry = 0). At the end of the last two rows, the counted numbers are shown as N_1 (the number of wet days) and N_0 (the number of dry days). The third column presents the event that the current day is wet with the previous wet day ($X_t = 1|X_{t-1} = 1$), and its corresponding probability (P_{11}) is calculated with the number of this event (N_{11}) over the total number of wet days, i.e.,

$$P_{11} = \frac{N_{11}}{N_1} = \frac{8}{14 - 1} = 0.62$$

Note that $N_1 - 1$ is employed, since the first day (X_1) is wet and its previous day (X_0) does not exist. The P_{01} value is

TABLE 6.1
Sample Daily Precipitation Data (mm) for the Month of July 2002 from Seoul, Korea

Precipitation	Occurrence	N_{11}	N_{01}
0.50	1.00		
5.00	1.00	1.00	
0.00	0.00		
0.00	0.00		
46.00	1.00		1.00
25.00	1.00	1.00	
0.00	0.00		
0.00	0.00		
0.00	0.00		
0.00	0.00		
0.00	0.00		
0.00	0.00		
2.50	1.00		1.00
22.00	1.00	1.00	
0.50	1.00	1.00	
0.00	0.00		
0.00	0.00		
0.00	0.00		
60.50	1.00		1.00
0.50	1.00	1.00	
0.00	0.00		
14.00	1.00		1.00
42.00	1.00	1.00	
0.10	1.00	1.00	
0.00	0.00		
0.00	0.00		
0.00	0.00		
0.00	0.00		
1.00	1.00		1.00
1.00	1.00	1.00	
0.00	0.00		
N_1	13	$N_{11} = 8$	$N_{01} = 5$
N_0	17	$P_{11} = 0.62$	$P_{01} = 0.29$

$$P_{01} = \frac{N_{01}}{N_0} = \frac{5}{17} = 0.29$$

The transition matrix is

$$\mathbf{P}_T = \begin{bmatrix} P_{00} & P_{01} \\ P_{10} & P_{11} \end{bmatrix} = \begin{bmatrix} 0.71 & 0.29 \\ 0.38 & 0.62 \end{bmatrix}$$

Also, the limiting probability can be calculated as

$$P_1 = \frac{N_1}{N} = \frac{14}{31} = 0.45 \quad \text{and} \quad P_0 = \frac{N_0}{N} = \frac{17}{31} = 0.55$$

or from Eq. (6.20)

$$P_1 = \frac{P_{01}}{P_{10} + P_{01}} = \frac{0.29}{0.38 + 0.29} = 0.43$$

Note that these two results of P_1 cannot agree with each other because the estimation is made with a limited number of data.

Example 6.2

Simulate the daily precipitation occurrence (i.e., wet or dry) with the estimated transition matrix in Example 6.1.

Solution:

1. At first, the occurrence of the first day (X_1) employs P_1. From simulating a uniform random number $r \sim Unif[0,1]$, the uniform random number u is compared with P_1. If $u < P_1$, then $X_1 = 1$, otherwise 0. Let $u = 0.5377$, and $u > P_1(=0.43)$. Therefore, $X_1 = 0$.
2. The occurrence of the following day (X_2) is simulated by employing the P_{01} probability, since the previous day is dry (i.e., $X_1 = 0$). Let the simulated uniform random number be $u = 0.127$. Since u $(=0.127) < P_{01}$ $(=0.29)$, $X_2 = 1$.
3. Repeat step (2) until all the occurrences are simulated.

6.2 WEATHER GENERATOR

One of the most commonly employed stochastic downscaling models is the Weather Generator (WGEN) developed by Richardson and Wright (1984). The major task of the WGEN model is to provide generated values of precipitation (Prcp), maximum temperature (T_{max}), minimum temperature (T_{min}), and solar radiation (SR) at a given location.

In the WGEN model, precipitation occurrence is modeled with the first-order Markov chain model. The occurrence probabilities are estimated for each month as in Eqs. (6.18) and (6.19). The amount of precipitation is simulated with a gamma distribution in Eq. (2.25) when precipitation occurs from the step given earlier that the parameters are constant in a given month. In other words, the parameters of the gamma distribution for the precipitation amount are estimated for each month. The three weather variables employed (maximum temperature, minimum temperature, and SR) are modeled with MAR(1) model of Eq. (6.3). The daily mean and variance of the three weather variables are fitted to a cosine function for taking the annual cycle of those statistics into account.

The parameter estimation of the WGEN model is separated into two parts as precipitation and other weather variables.

6.2.1 Model Fitting

6.2.1.1 Precipitation

The parameter estimation from precipitation should be done for each month. Four parameters must be estimated from daily precipitation data of each month as P_{01} and P_{11} with Eq. (6.19) for the occurrence and α and β with the method of moments in Eqs. (2.30) and (2.29) or with Maximum Likelihood Estimation (MLE) in Eqs. (2.43) and (2.42), respectively, for the precipitation amount.

6.2.1.2 Weather Variables (T_{max}, T_{min}, SR)

a. The mean and standard deviation for each day (d = 1,..., 365) of a year, denoted as $\hat{\mu}_{s|x_t=a_t}(d)$ and $\hat{\sigma}_{s|x_t=a_t}(d)$, are estimated with the wet or dry condition of the precipitation data. Here, $\hat{\mu}_{s|x_t=a_t}(d)$ and $\hat{\sigma}_{s|x_t=a_t}(d)$ show the mean and standard deviation of the sth variable for the day d (=1,...,365) when the precipitation condition (x_t) is at a_t (=0 or 1). Here, s = 1 represents T_{max} while s = 2 and 3 do T_{min} and SR, respectively.
b. Each statistic with the wet or dry condition (here, denoted as u_d) is fitted to a cosine function, such that

$$u_d = \bar{u} + C\cos\left[0.0172(d-T)\right], d = 1,\ldots,365 \tag{6.21}$$

where $\bar{u}$ is the mean of u_d; C is the amplitude of the harmonic; and T is the offset of harmonic in days.
c. The standardized data of the three weather variables is estimated with

$$y_t^s = \frac{z_t^s - \hat{\mu}_{s|x_t=a_t}(d)}{\hat{\sigma}_{s|x_t=a_t}(d)} \tag{6.22}$$

d. The parameter matrices **A** and **B** for the MAR(1) model in Eqs. (6.9) and (6.10) are estimated with the covariance matrix and lag-one covariance matrix, $\hat{\mathbf{M}}_0$ and $\hat{\mathbf{M}}_1$ of all three standardized data y_t^s for all record lengths (i.e., t = 1,...,n_{yr} × 365 and s = 1,2,3, here and n_{yr} is the number of observation record years).

Example 6.3

Fit the WGEN model for four variables (maximum temperature, minimum temperature, SR, and precipitation) to the daily weather dataset of Seoul, Korea for 30 years (1981–2010) and the December 31 data of 30 years is shown in Table 6.2 as an example.

PRECIPITATION

Solution:

The four parameters (P_{01}, P_{11}, α, and β) of daily precipitation are estimated for each month. As an example, Table 6.1 and Example 6.1 illustrate how P_{01} and P_{11} with Eq. (6.19) are estimated. The parameters of the gamma distribution, α and β, are

TABLE 6.2
Sample Weather Data for the Last Day of a Year (December 31) from Seoul, Korea

Years	T_{max} (°C)	T_{min} (°C)	SR (MJ/m²)	Precipitation (mm)
1981	1.20	−4.90	2.82	0.40
1982	2.80	−5.00	2.91	0.00
1983	−2.10	−10.10	6.13	0.00
1984	1.40	−4.50	4.47	1.10
1985	−1.10	−7.20	8.29	0.00
1986	4.50	−6.00	6.23	1.80
1987	0.00	−10.60	7.14	0.00
1988	2.10	−3.30	1.56	0.00
1989	0.10	−8.70	6.61	0.00
1990	8.80	−3.40	6.00	0.00
1991	4.10	−3.70	4.88	0.00
1992	2.60	−3.70	6.65	0.00
1993	1.30	−4.20	5.00	0.00
1994	1.30	−4.00	8.44	0.00
1995	1.80	−7.30	2.95	0.00
1996	7.00	0.30	3.26	0.00
1997	5.70	−3.90	6.94	0.00
1998	1.30	−6.00	7.85	0.00
1999	7.30	0.00	3.33	0.00
2000	2.30	−8.20	7.76	0.00
2001	1.30	−6.80	3.64	6.70
2002	1.90	−3.70	8.10	0.10
2003	5.80	−0.30	3.65	0.00
2004	−1.20	−7.00	8.07	0.00
2005	5.50	−2.00	5.96	0.00
2006	6.10	−4.40	8.54	0.00
2007	−2.90	−7.00	8.40	0.00
2008	−2.90	−8.00	9.21	0.00
2009	−8.30	−12.90	10.27	0.00
2010	−6.20	−10.70	10.04	0.00
μ_{Dry}	1.65	−5.65	6.39	0.00
μ_{Wet}	2.06	−5.18	5.05	2.02
σ_{Dry}	4.21	3.48	2.41	0.00
σ_{Wet}	1.39	1.23	2.12	2.70

estimated with the mean and variance of the precipitation amount for rainy days. If the mean and variance are 16.89 and 430.73, respectively, then the parameters are $\beta = 430.73/16.89 = 25.50$ and $\alpha = 16.89/25.50 = 0.66$. See Eqs. (2.29) and (2.30) for further detail.

The estimated monthly statistics for Seoul with the 30-year dataset are shown in Figure 6.1 for P_{01} and P_{11} as well as in Figure 6.2 for the gamma parameters

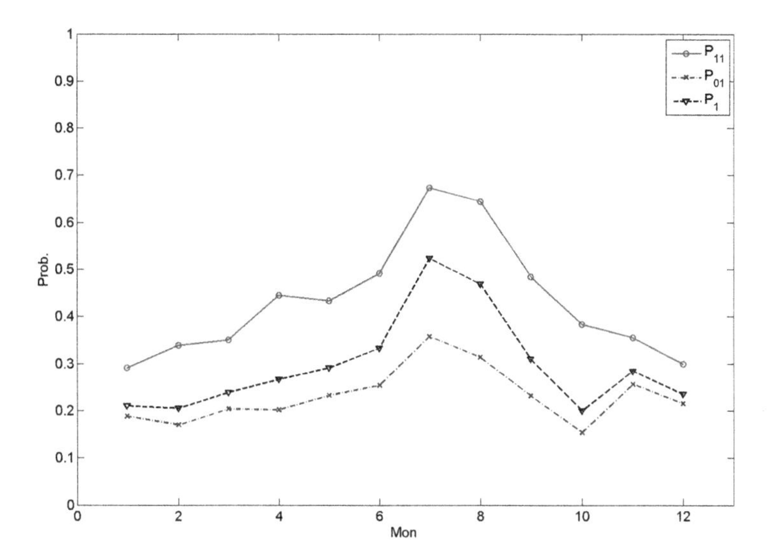

FIGURE 6.1 Transition probabilities (P_{11} and P_{01}) and limiting probability (P_1) for each month for 30 years (1981–2010) of daily precipitation data from Seoul, Korea. See Table 6.3 for detailed values.

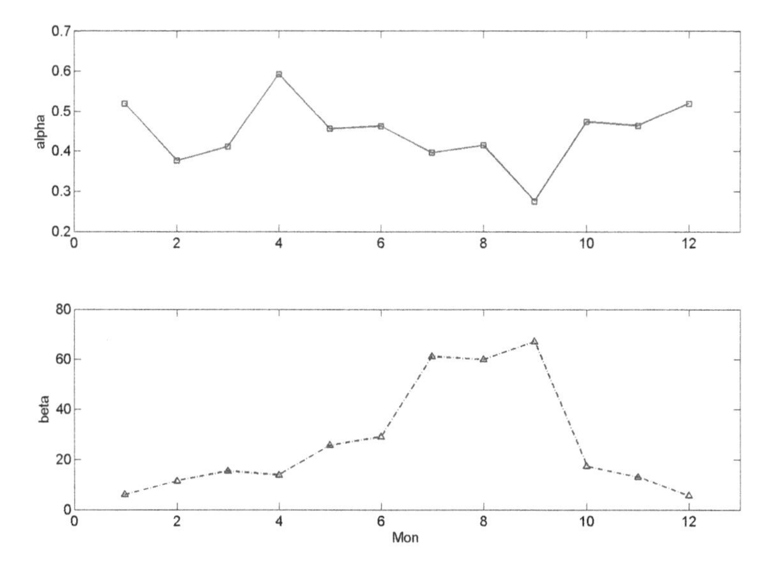

FIGURE 6.2 Parameters of the fitted gamma distribution (α and β) for each month for 30 years (1981–2010) of daily precipitation data from Seoul, Korea.

α, β. Note that high probability of precipitation occurrence is evident during the summer season in Figure 6.1. In Figure 6.2, the shape parameter of the gamma distribution is always smaller than 1.0, indicating that the fitted gamma distribution is exponentially decreasing. The scale parameter presents the seasonal cycle similar to the occurrence probabilities in Figure 6.1.

TABLE 6.3
Parameters Estimated for the Precipitation Data from Seoul for Observation, Control, and Future Scenario of RCP8.5 and for the Adjusted Parameter for RCP8.5 Future Scenario

	P_{11}	P_{01}	α	β	P_{11}	P_{01}	α	β
Mon			**Observation**				**Adjust (RCP8.5)**	
1	0.29	0.19	0.52	6.12	0.33	0.15	0.26	20.03
2	0.34	0.17	0.38	11.51	0.32	0.18	0.79	5.68
3	0.35	0.20	0.41	15.41	0.34	0.17	0.43	18.42
4	0.44	0.20	0.59	13.88	0.41	0.22	0.62	12.28
5	0.43	0.23	0.46	25.71	0.51	0.25	0.63	19.73
6	0.49	0.25	0.46	29.09	0.47	0.27	0.41	30.81
7	0.67	0.36	0.40	61.25	0.64	0.35	0.30	72.52
8	0.64	0.31	0.41	60.01	0.64	0.30	0.64	39.54
9	0.48	0.23	0.28	67.23	0.50	0.22	0.17	102.24
10	0.38	0.15	0.47	17.43	0.43	0.14	0.53	17.67
11	0.36	0.26	0.46	13.06	0.34	0.19	0.56	10.62
12	0.30	0.22	0.52	5.58	0.34	0.25	0.37	7.57
Mon			**Control**				**Future (RCP8.5)**	
1	0.21	0.11	0.65	5.51	0.26	0.08	0.32	18.04
2	0.24	0.11	0.19	30.47	0.22	0.12	0.40	15.03
3	0.38	0.22	0.39	18.02	0.37	0.18	0.41	21.54
4	0.48	0.22	0.37	39.58	0.45	0.24	0.38	35.02
5	0.44	0.22	0.21	61.62	0.51	0.23	0.29	47.29
6	0.64	0.26	0.30	68.45	0.62	0.28	0.27	72.48
7	0.78	0.29	0.42	55.21	0.75	0.28	0.32	65.37
8	0.65	0.26	0.22	52.36	0.65	0.25	0.35	34.50
9	0.54	0.15	0.29	46.38	0.55	0.14	0.18	70.53
10	0.35	0.19	0.22	43.27	0.40	0.18	0.24	43.85
11	0.33	0.23	0.35	26.01	0.32	0.17	0.42	21.16
12	0.26	0.10	0.36	19.09	0.29	0.12	0.26	25.90

WEATHER VARIABLE

1. Estimate the mean and variance of three variables for each day (not each month) according to the condition of the precipitation occurrence of the present day (i.e., $\hat{\mu}_{s|X_t=a_t}(d)$ and $\hat{\sigma}_{s|X_t=a_t}(d)$). The estimated parameters of mean and standard deviation are shown in Figure 6.3 and Table 6.4 as well as their fitted cosine function (overlapped thick lines). Note that the temperature variables present the opposite convex between mean and standard deviation, while the SR shows the same upward convex for these two statistics. T_{max} presents a higher difference than T_{min} between wet and dry conditions. Obviously, the T_{max} mean for the wet condition (see the dash-dotted line at the top left panel of Figure 6.3 and Table 6.4) is lower than the one for the dry condition (the dash-dotted line) during

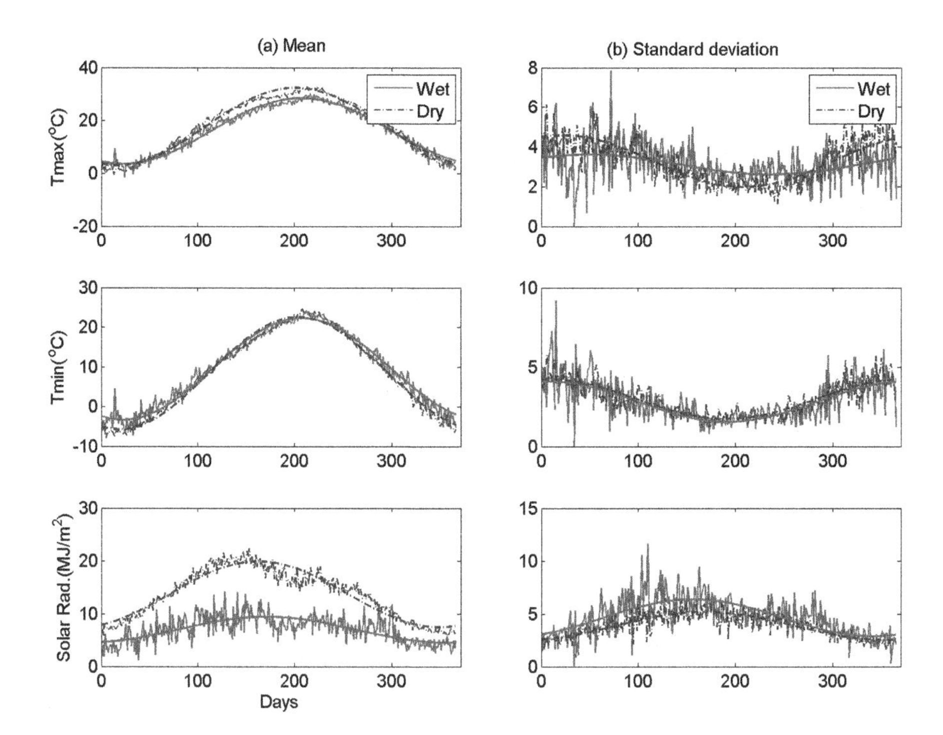

FIGURE 6.3 (See color insert.) Mean (left column panels) and standard deviation (right column panels) of three weather variables (T_{max}, T_{min}, SR) for each day ($d = 1,\ldots,365$) regarding their occurrence condition (wet: solid line or dry: dash-dotted line) for Seoul as well as their fitted cosine function [thick smoothed lines overlapping with the original estimations from the data, see Eq. (6.21)]. Note that the parameters of the fitted cosine function in Eq. (6.21) were estimated with the Harmony Search algorithm presented in Chapter 2.

the summer season around 200 days, because precipitation absorbs the heat resulting in the decrease of air temperature and vice versa for minimum temperature during the winter season (see the middle left panel) around 1–50 and 300–365 days.

2. Obtain the standardized daily weather variables employing Eq. (6.22). Note that the mean and standard deviation employed for standardization are the ones directly estimated from the data and not from smoothed ones.
3. Calculate the correlation and lag-one correlation matrices ($\hat{\mathbf{M}}_0$ and $\hat{\mathbf{M}}_1$) with the standardized daily weather data and estimate the parameter matrices of **A** and **B** for the MAR(1) model. Note that all the data values are employed in calculation. The estimated $\hat{\mathbf{M}}_0$ and $\hat{\mathbf{M}}_1$ are

$$\hat{\mathbf{M}}_0 = \begin{bmatrix} 1.0 & 0.696 & 0.072 \\ 0.696 & 1.0 & -0.261 \\ 0.072 & -0.261 & 1.0 \end{bmatrix}$$

and

TABLE 6.4

Estimated Mean and Standard Deviation of the T_{max} Weather Variable Related to the Precipitation Occurrence and Its Cosine-Fitted Smoothing Values

	Mean				Stand. Dev.			
	Original		Cosine Fitted		Original		Cosine Fitted	
Day	Wet	Dry	Wet	Dry	Wet	Dry	Wet	Dry
1	2.84	1.56	4.34	3.49	3.54	3.66	3.45	4.49
2	1.76	1.85	4.26	3.43	4.56	3.80	3.46	4.50
3	1.47	2.37	4.19	3.37	2.21	4.81	3.47	4.51
4	0.64	1.12	4.13	3.31	5.61	4.82	3.47	4.51
5	3.40	−0.15	4.06	3.26	1.90	6.12	3.48	4.52
6	3.09	1.27	4.01	3.22	3.64	5.08	3.48	4.53
7	2.88	1.07	3.95	3.18	4.20	3.75	3.49	4.53
8	3.62	1.31	3.90	3.14	3.57	4.17	3.49	4.54
	⋮		⋮		⋮		⋮	
200	26.14	30.92	28.38	32.39	2.73	1.64	2.70	1.99
201	27.21	31.47	28.41	32.39	2.72	1.92	2.69	1.99
202	27.61	30.45	28.43	32.39	2.29	2.13	2.69	1.99
203	26.89	31.62	28.44	32.38	1.60	1.92	2.69	1.99
204	27.06	31.13	28.46	32.37	2.03	2.64	2.68	1.99
205	28.21	30.25	28.47	32.35	2.80	3.13	2.68	1.98
206	27.87	30.99	28.47	32.34	2.37	2.80	2.67	1.98
207	28.47	31.55	28.47	32.31	3.02	3.00	2.67	1.98
	⋮		⋮		⋮		⋮	
358	1.99	4.21	5.54	4.61	3.67	5.10	3.37	4.36
359	2.17	3.93	5.43	4.50	2.10	5.44	3.37	4.38
360	4.25	2.25	5.32	4.39	1.34	5.82	3.38	4.39
361	2.80	2.25	5.21	4.29	4.82	4.47	3.39	4.40
362	3.86	2.17	5.11	4.19	4.00	4.88	3.39	4.41
363	2.87	1.79	5.00	4.09	3.24	4.35	3.40	4.42
364	2.55	1.35	4.91	4.00	3.58	3.45	3.41	4.43
365	2.06	1.65	4.81	3.91	1.39	4.21	3.42	4.44

$$\hat{\mathbf{M}}_1 = \begin{bmatrix} 0.567 & 0.488 & -0.063 \\ 0.630 & 0.662 & -0.148 \\ -0.129 & -0.137 & 0.255 \end{bmatrix}$$

The parameter matrices $\hat{\mathbf{A}}$ and $\hat{\mathbf{B}}$ from Eqs. (6.9) and (6.10) with Eq. (6.16) are estimated as

$$\hat{\mathbf{A}} = \begin{bmatrix} 0.469 & 0.147 & -0.058 \\ 0.365 & 0.389 & -0.073 \\ -0.209 & 0.085 & 0.293 \end{bmatrix}$$

and

$$\hat{\mathbf{B}} = \begin{bmatrix} 0.812 & 0 & 0 \\ 0.363 & 0.609 & 0 \\ 0.206 & -0.357 & 0.861 \end{bmatrix}$$

6.2.2 Simulation of Weather Variables

From the fitted WGEN model, it is straightforward to simulate weather variables with precipitation first and the other variables with their simulated occurrence condition.

6.2.2.1 Precipitation

a. The occurrence of daily precipitation must be simulated with the transition matrix $\mathbf{P}_T$ in Eq. (6.18) at first. Note that this transition matrix varies for each month. For the first day (i.e., $t = 1$), no previous condition (X_0) is available. There are two ways: (1) simulate the occurrence with the limiting probability (P_1) or (2) randomly assign 0 or 1 and remove the first simulated year in the total simulation years as a warming period.
b. The precipitation amount is simulated from a gamma distribution when it is a rainy day (i.e., $X_t = 1$) from the estimated parameters (α and β) for each month.

6.2.2.2 Weather Variables (T_{max}, T_{min}, SR)

a. Simulate three standardized weather variables employing the MAR(1) model with the estimated parameters **A** and **B** (Y_t^s, $s = 1, 2, 3$ and $t = 1,\ldots,N^G \times 365$, and N^G is the number of simulation years).
b. Assign the appropriate mean and standard deviation to the simulated standardized variables earlier as

$$Z_t^s = \hat{\sigma}_{s|X_t}(d) Y_t^s + \hat{\mu}_{s|X_t}(d) \tag{6.23}$$

where $\hat{\mu}_{s|X_t}(d)$ is defined according to the wet or dry condition of the simulated precipitation occurrence (X_t). Instead of employing the standard deviation $\hat{\sigma}_{s|X_t}$, the coefficient of variation can be used as

$$Z_t^s = \hat{\mu}_{s|X_t}(d)\left[\hat{\eta}_{s|X_t}(d) Y_t^s + 1\right] \tag{6.24}$$

where $\hat{\eta}_s(d)$ represents the coefficient of variation.

Example 6.4

Generate four weather variables (maximum temperature, minimum temperature, SR, and precipitation) with the fitted WGEN model in Example 6.3.

PRECIPITATION

Solution:

A. Occurrence

The occurrence of precipitation is simulated with the estimated parameters.

a. The first day X_1 is simulated by employing P_1 of the first month [i.e., $P_1 = 0.21$ as in Table 6.3 and with Eq. (6.20)]. Generate a uniform random number as in Section 2.3, say $r = 0.535$, and the first day is = dry (i.e., $X_1 = 0$ since $r > P_1$ as $0.535 > 0.21$).
b. The occurrence of the second day is simulated with the probability P_{01} of the first month (i.e., $P_{01} = 0.19$). Simulate a uniform random number, say $r = 0.132$. Then the second day is wet (i.e., $X_2 = 1$ since $r < P_{01}$ as $0.132 < 0.19$).
c. Continue this simulation by employing an appropriate transition probability for the month. For example, if $t = 32$ and $X_{31} = 1$, then P_{11} of the second month is employed (i.e., $P_{11} = 0.34$). The same strategy is followed even for the following year. For example, if $t = 366$ and $X_{365} = 0$, then P_{01} of the first month is employed (i.e., $P_{01} = 0.19$).
d. Repeat this until all the target occurrences are generated.

B. Amount

The precipitation amount is simulated whenever a day is wet (i.e., $X_t = 1$) from the occurrence simulation from the fitted gamma distribution with the parameter set for the corresponding month. For example, since the second day is wet ($X_2 = 1$) as the occurrence example given earlier, the precipitation amount is generated from the gamma function with $\alpha = 0.52$ and $\beta = 6.116$.

WEATHER VARIABLES

A. MAR(1)

Three standard normal variables $\mathbf{Y}_t$ are simulated through the fitted MAR(1) model with the parameters **A** and **B**.

a. The first day is simulated with $\hat{\mathbf{B}}\mathbf{E}_t$ in Eq. (6.3), since there is no previous condition ($\mathbf{Y}_{t-1}$) available. Assume three values are simulated from the standard normal distribution, say $\mathbf{E}_1 = [-1.303, 0.795, 1.802]^T$. Then

$$\mathbf{Y}_1 = \hat{\mathbf{B}}\mathbf{E}_1 = \begin{bmatrix} 0.812 & 0 & 0 \\ 0.363 & 0.609 & 0 \\ 0.206 & -0.357 & 0.861 \end{bmatrix} \begin{bmatrix} -1.303 \\ 0.795 \\ 1.802 \end{bmatrix} = \begin{bmatrix} -1.058 \\ 0.011 \\ 0.999 \end{bmatrix}$$

b. The next day is simulated with this condition ($\mathbf{Y}_1$). Say $\mathbf{E}_2 = [0.023, -1.915, 1.192]^T$, then

$$\mathbf{Y}_2 = \hat{\mathbf{A}}_1\mathbf{Y}_1 + \hat{\mathbf{B}}\mathbf{E}_2 = \begin{bmatrix} 0.469 & 0.147 & -0.058 \\ 0.365 & 0.389 & -0.073 \\ -0.209 & 0.085 & 0.293 \end{bmatrix} \begin{bmatrix} -1.058 \\ 0.011 \\ 0.999 \end{bmatrix} +$$

$$\begin{bmatrix} 0.812 & 0 & 0 \\ 0.363 & 0.609 & 0 \\ 0.206 & -0.357 & 0.861 \end{bmatrix} \begin{bmatrix} 0.023 \\ -1.915 \\ 1.192 \end{bmatrix} = \begin{bmatrix} -0.534 \\ -1.611 \\ 1.193 \end{bmatrix}$$

c. Continue this simulation until $t = N_C \times 365$, where N_C is the number of simulation years.

B. Back-standardization

The simulated standard normal variables from the earlier MAR(1) model are back-standardized by adding mean and multiplying by standard deviation according to the condition of precipitation occurrence of the current day (X_t) as in Eq. (6.23). The estimated mean and standard deviation of T_{max} for each day according to the wet or dry condition are shown at the left side of Table 6.4 as well as its cosine-fitted smoothing values (right side of Table 6.4). In this back-standardization, the cosine-fitted statistics are employed.

As mentioned earlier, the first and second simulated days are dry and wet, respectively [$X_1 = 0$, $X_2 = 1$]. Therefore,

$$Z_1^1 = \hat{\sigma}_{1|X_1=0}(1)Y_1^1 + \hat{\mu}_{1|X_1=0}(1) = 4.49 \times -1.058 + 3.49 = -1.26°\text{C}$$

$$Z_2^1 = \hat{\sigma}_{2|X_2=1}(2)Y_2^1 + \hat{\mu}_{2|X_2=1}(2) = 3.46 \times -0.534 + 4.26 = 6.11°\text{C}$$

6.2.3 Implementation of Downscaling

The downscaling of GCM with the WGEN model can be done by changing the parameters of the model. Wilby et al. (1998) suggested a way to implement the statistical downscaling in the WGEN model, especially for the precipitation variable.

The transition probabilities of the Markov chain model are characterized by the wet-day probability (P_1) with the transition probability as in Eq. (6.20) and lag-one autocorrelation (γ) can be described with the transition probabilities as

$$\gamma = P_{11} - P_{01} \tag{6.25}$$

The adjusted wet-day probability parameter employing the logit function (inverse of the logistic function) is

$$P_1^{adj} = \Psi^{-1}\left\{\Psi\left(P_1^O\right) + \Psi\left(P_1^F\right) - \Psi\left(P_1^C\right)\right\} \tag{6.26}$$

Here, the standard logistic function is $f(x) = \exp(x)/(\exp(x)+1)$ and its inverse $\Psi(z) = \ln\left(z/(1-z)\right)$ and superscripts O, F, and C indicate "observation," "future scenario," and "controlled scenario," respectively.

The correlation γ can be adjusted as

$$\gamma^{adj} = \Theta^{-1}\left\{\Theta\left(\gamma^O\right) + \Theta\left(\gamma^F\right) - \Theta\left(\gamma^C\right)\right\} \tag{6.27}$$

where $\Theta(z) = 0.5\ \ln\left[(1+z)/(1-z)\right]$ and $\Theta^{-1}(x) = \left\{(\exp(2x)-1)/(\exp(2x)+1)\right\} = \tanh(x)$.

From these two adjusted statistics, the key transition probabilities are estimated as

$$P_{01}^{adj} = \left(1-\gamma^{adj}\right)P_1^{adj} \tag{6.28}$$

$$P_{11}^{adj} = \gamma^{adj} + P_{01}^{adj} \tag{6.29}$$

The parameters of the gamma distribution can be adjusted as

$$\alpha^{adj} = \frac{\alpha^O \alpha^F}{\alpha^C} \tag{6.30}$$

$$\beta^{adj} = \frac{\beta^O \beta^F}{\beta^C} \tag{6.31}$$

For the other weather variables, their mean and standard deviation as in Eq. (6.22) can be adjusted according to changes, i.e.,

$$\mu^{adj} = \mu^O + \left(\mu^F - \mu^C\right) \tag{6.32}$$

$$\sigma^{adj} = \frac{\sigma^O \sigma^F}{\sigma^C} \tag{6.33}$$

while parameter matrices $\mathbf{A}_1$ and $\mathbf{B}$ for the MAR(1) model are generally fixed.

Example 6.5

Adjust the future scenario of RCP8.5 in Example 6.3. The estimated parameters from the controlled run and the future RCP scenario of 2006–2100 are shown in the lower part of Table 6.3 that is obtained from HadGEM3-RA (control run: 1979–2005 and future scenario RCP8.5 2006–2100).

OCCURRENCE

Solution:

For the first month, the correlation and P_1 are

$$\gamma^O = P_{11}^O - P_{01}^O = 0.29 - 0.19 = 0.1$$

$$P_1^O = \frac{P_{01}^O}{\left(1+P_{01}^O - P_{11}^O\right)} = \frac{0.19}{(1+0.19-0.29)} = 0.21$$

and $\left[P_1^C, P_1^F\right] = [0.12, 0.1]$ and $\left[\gamma_1^C, \gamma_1^F\right] = [0.10, 0.18]$.

$$P_1^{adj} = \Psi^{-1}\left\{\Psi\left(P_1^O\right) + \Psi\left(P_1^F\right) - \Psi\left(P_1^C\right)\right\} = \Psi^{-1}\{-1.32 - 2.18 + 2.01\} = 0.183$$

$$\gamma^{adj} = \Theta^{-1}\left\{\Theta\left(\gamma^O\right) + \Theta\left(\gamma^F\right) - \Theta\left(\gamma^C\right)\right\} = \Theta^{-1}\{0.10 + 0.18 - 0.10\} = 0.180$$

Therefore,

$$P_{01}^{adj} = \left(1-\gamma^{adj}\right)P_{1}^{adj} = (1-0.180)0.183 = 0.15$$

$$P_{11}^{adj} = \gamma^{adj} + P_{01}^{adj} = 0.18 + 0.15 = 0.33$$

Amount

$$\alpha^{adj} = \frac{\alpha^{O}\alpha^{F}}{\alpha^{C}} = 0.52\frac{0.32}{0.65} = 0.26$$

$$\beta^{adj} = \frac{\beta^{O}\beta^{F}}{\beta^{C}} = 6.12\frac{18.04}{5.11} = 20.03$$

All the parameters for precipitation are shown in the right-upper part of Table 6.3 and Figure 6.4. The simulation can be made as in Example 6.4 with these adjusted parameters.

6.3 NONPARAMETRIC WEATHER GENERATOR

Nonparametric WGEN (NWG) had been designed in the 1990s (Young, 1994; Rajagopalan and Lall, 1999) with K-nearest neighbor bootstrap method for

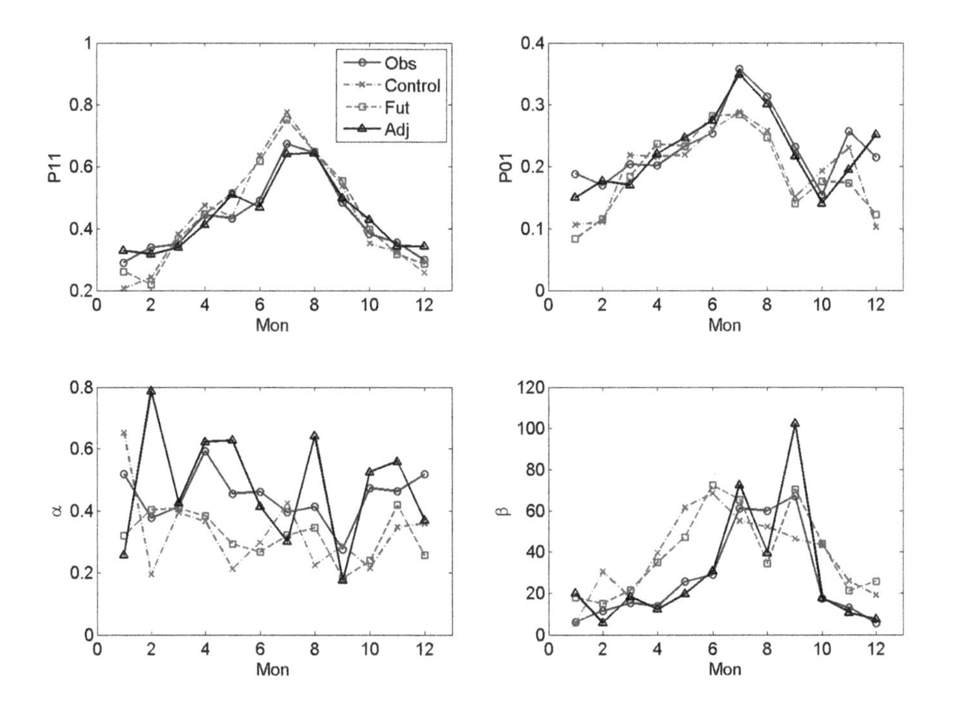

FIGURE 6.4 (See color insert.) Adjusted parameters of the WGEN model for the RCP8.5 future scenario from the HadGem3-RA developed by Korea Meteorological Administration. Control run implies the HadGem3-RA dataset of 1979–2005, while the future scenario is 2006–2100.

time-series resampling to daily weather data. Lee et al. (2012) modified the NWG model by applying the genetic algorithm and gamma kernel density to produce new weather patterns and values. Yates et al. (2003) updated this scheme to generate alternative climate scenarios and further improvement by Sharif and Burn (2006). Here, the NWG is explained under the scheme of Yates et al. (2003) and Sharif and Burn (2006).

6.3.1 Simulation under Current Climate

In NWG, the K-Nearest Neighbor Resampling (KNNR) algorithm is generally employed with the Mahalanobis distance. Here, S number of daily weather variables (e.g., maximum temperature: T_{max}, minimum temperature: T_{min}, solar radiation: SR, and precipitation: PT) are considered as aforementioned with n number of data. Let the weather variables be generated at days $t+1=1,\ldots,365$ for all the generation year, $ig=1,\ldots,N^G$. The following procedure can be used:

1. Set $\mathbf{x}_{iy,d}=\left[x_{iy,d}^1,x_{iy,d}^2,\ldots,x_{iy,d}^S\right]^T$ and $iy=1,\ldots,n_{yr}$; and $d=1,\ldots,365$, where n_{yr} is the number of observed years.
2. Select all days within the temporal window of width b centered on the current day t as potential candidates for day $t+1$. For example, if a 14-day window is selected ($b=14$) and $t=1$ (Jan. 1), then front and back of the days centered on January 1 for all n_{yr} are included, i.e., December 25 and January 8. There are $[(b+1)\times n_{yr}]-1$ days for potential neighbors.
3. Compute the covariance matrix $\mathbf{C}$ with the selected neighbors from step (2).
4. For $ig=1$ and $t=1$, generate randomly all four weather variables among observational data, i.e., $iy=1,\ldots,n_{yr}$ and $d=1$.
5. Let $\mathbf{X}_{ig,t}=\left[X_{ig,t}^1,X_{ig,t}^2,\ldots,X_{ig,t}^S\right]^T$ be known from the previous generation.
6. Estimate the Mahalanobis distances for all $[(b+1)\times n_{yr}]-1$ days as follows:

$$D_j=\left(\mathbf{X}_{ig,t}-\mathbf{x}_j\right)^T\mathbf{C}^{-1}\left(\mathbf{X}_{ig,t}-\mathbf{x}_j\right)^T \quad j=1,\ldots,\left[(b+1)\times n_{yr}\right]-1 \qquad (6.34)$$

7. Sort the estimated distances D_j from the smallest to the largest, and the smallest K neighbors are retained. Heuristic choice $K=Integer\left(\sqrt{(b+1)n_{yr}-1}\right)$ can be easily adapted to select K (Lall and Sharma, 1996).
8. Randomly select one of the stored k time indices with the weighting probability as in Eq. (8.2).

$$w_m=\frac{\frac{1}{m}}{\sum_{j=1}^{k}\frac{1}{j}} \quad m=1,\ldots,k,$$

9. Assign the proceeding weather values of the selected index for $\mathbf{X}_{ig,t+1}$. For example, if $iy = 10$ for $id = 2$ is selected, then $\mathbf{X}_{ig,t+1} = \mathbf{x}_{10,3} = \left[x_{10,3}^1, x_{10,3}^2, \ldots, x_{10,3}^S\right]^T$.
10. Repeat steps of (5)–(9) until all the required generation is done.

Example 6.6

Generate three weather variables (maximum temperature, minimum temperature, and precipitation) with the procedure of the NWG model explained earlier. Employ the observed dataset from the Seoul station in South Korea for ten years from 1981 to 1990 in Table 6.5.

Solution:

The window width (b) of 14 is normally used, but here $b = 4$ is applied for simplicity. From the dataset, $n_{yr} = 10$ and assume that the target day is July 13 (194th day) and $ig = 1$. Its previous simulation data (i.e., $ig = 1$ and $t = 193$) is known as $\mathbf{X}_{ig,t} = [28.8, 24.5, 14.0]^T$.

TABLE 6.5
Dataset of Weather Variables for NWG

Ind	Year	M	D	T_{max}	T_{min}	Prcp	D_j	Ind	Year	M	D	T_{max}	T_{min}	Prcp	D_j
1	1981	7	11	28.8	18.8	21.2	19.6	26	1986	7	11	23.3	19.7	52.5	19.3
2	1981	7	12	24.3	20.5	82.5	21.2	27	1986	7	12	27.1	19.2	4.7	17.7
3	1981	7	13	29.3	20.3	1.0	10.3	28	1986	7	13	25.3	19.6	1.3	18.9
4	1981	7	14	29.8	21.7	0.0	4.7	29	1986	7	14	28.5	20.5	15.6	9.4
5	1981	7	15	30.6	22.9	0.0	1.9	30	1986	7	15	29.3	21	6.1	7.1
6	1982	7	11	28.8	22.4	7.8	2.6	31	1987	7	11	24.5	21	61.3	13.0
7	1982	7	12	28.1	20.2	0.0	11.6	32	1987	7	12	25.3	20.9	0	12.9
8	1982	7	13	28.3	19.7	0.0	14.0	33	1987	7	13	26.8	20.3	0	12.6
9	1982	7	14	33.7	18.5	0.0	24.9	34	1987	7	14	25.3	19.8	0	18.1
10	1982	7	15	30.9	17.4	7.7	30.1	35	1987	7	15	23.1	20.2	41.7	16.3
11	1983	7	11	29.3	19.7	0.0	13.5	36	1988	7	11	27.5	21.6	0	6.4
12	1983	7	12	28.4	18.8	0.0	19.4	37	1988	7	12	26.2	22.2	52.2	6.8
13	1983	7	13	29.5	19.1	0.0	17.0	38	1988	7	13	28.8	24.5	14	0.0
14	1983	7	14	22.5	19.0	58.3	25.4	39	1988	7	14	25.4	23.2	44.6	3.4
15	1983	7	15	27.8	20.0	0.0	12.8	40	1988	7	15	30	22.2	1	3.2
16	1984	7	11	29.4	20.9	0.0	7.7	41	1989	7	11	23.6	19.8	51.8	18.3
17	1984	7	12	23.9	20.7	63.7	15.1	42	1989	7	12	28.6	19.7	0	13.8
18	1984	7	13	26.7	21.4	0.6	8.1	43	1989	7	13	29.6	19.6	0	14.0
19	1984	7	14	30.1	20.9	0.1	7.7	44	1989	7	14	31	19.6	0	14.4
20	1984	7	15	32.7	21.0	14.7	11.4	45	1989	7	15	25.8	20.3	1.5	14.2
21	1985	7	11	28.1	21.2	0.1	7.2	46	1990	7	11	26.2	22.4	39	4.3
22	1985	7	12	23.5	21.6	25.5	10.3	47	1990	7	12	31.4	22	9.4	5.0
23	1985	7	13	27.3	21.3	0.2	7.7	48	1990	7	13	27.3	20.9	0.4	9.2
24	1985	7	14	29.3	20.3	0.1	10.4	49	1990	7	14	24	21.6	33	8.8
25	1985	7	15	30.7	20.2	0.0	11.0	50	1990	7	15	27.7	21.3	3.6	6.9

1. "K" can be estimated as $K = Integer\left(\sqrt{(2+1)10-1}\right) = 5$
2. The covariance matrix **C** and its inverse are calculated as

$$C = \begin{bmatrix} 6.9 & -0.2 & -38.9 \\ -0.2 & 1.7 & 3.1 \\ -38.9 & 3.1 & 495.5 \end{bmatrix} \text{and } C^{-1} = \begin{bmatrix} 0.257 & -0.006 & 0.020 \\ -0.006 & 0.587 & -0.004 \\ 0.020 & -0.004 & 0.004 \end{bmatrix}$$

3. With the previous dataset $\mathbf{X}_{ig,t} = [28.8, 24.5, 14.0]^T$, estimate the Mahalanobis distances between $\mathbf{X}_{ig,t}$ and the observed data in Table 6.5. For example, the first one (1981/7/11), $\mathbf{X}_{1,191} = [28.8, 18.8, 21.2]^T$ with

$$\mathbf{X}_{ig,t} - \mathbf{X}_{1,191} = [28.8, 24.5, 14.0]^T - [28.8, 18.8, 21.2]^T = [0, 5.7, -7.2]^T$$

$$D_1 = [0, 5.7, -7.2] \begin{bmatrix} 0.257 & -0.006 & 0.020 \\ -0.006 & 0.587 & -0.004 \\ 0.020 & -0.004 & 0.004 \end{bmatrix} \begin{bmatrix} 0 \\ 5.7 \\ -7.2 \end{bmatrix} = 19.6$$

All the estimated distances are shown in the seventh and last columns.
4. Select the smallest k (=5) distances. Here, the five smallest distances are 0 on 1988/7/13 (Ind: 38), 1.9 on 1981/7/15 (5), 2.6 at 1982/7/11 (6), 3.2 on 1988/7/15 (40), and 3.4 on 1988/7/14 (39).
5. Among five, one is selected with the weight probability of w_m, $m = 1,\ldots,5$.

$$w_1 = \frac{\frac{1}{1}}{\sum_{j=1}^{5}\frac{1}{j}} = \frac{\frac{1}{1}}{2.283} = 0.44$$

Also, $w_2 = 0.22$, $w_3 = 0.14$, $w_4 = 0.11$, and $w_5 = 0.09$. The selection among five is done by generating a uniform random number (say $u = 0.92$), and its index is obtained from the cumulative sum of w [WS = [0 0.44 0.66 0.80 0.91 1.00]] and WS is the range that its index is taken between the two values. For example, since $u = 0.92$, the fifth is selected.
6. The corresponding date for the selected fifth distance (D_j = 3.4) is 1988/7/14(39). Here, the proceeding weather values of the selected index 1988/7/15(40) for $\mathbf{X}_{ig,t+1} = \mathbf{x}_{40} = [30, 22.2, 44.6]$.
7. Repeat steps (3)–(6) until all the required simulation is completed. Note that the first value (i.e., $ig = 1$, $t = 1$) can be simulated by randomly selecting from the observed dataset of January 1 and its window to start with. The window can be extended to the previous year or the preceding year. For $t = 2$ (January 2) and $b = 4$ as an example, its window is extended to December 31 and December 30 of the previous year as well as January 3 and January 4 of the current year.

6.3.2 Simulation under Future Climate Scenarios

Yates et al. (2003) suggested a technique for simulating future climate scenarios using the KNNR algorithm. The basic idea is to collect the subset of the years with biasing the raked list (e.g., the years that has higher mean temperature or drier precipitation) with the integer function of

$$I_W^{i,N} = INT\left[N\left(1 - u^{\lambda_w^i}\right)\right] + 1$$

where λ_w^i is the shape parameter function. For example $\lambda_w^i = 1.0$ gives the equal probability that any year for the given week w. The main drawback of this technique is that a magnitude (such as a certain amount of increase or decrease temperature) cannot be specified.

6.4 SUMMARY AND CONCLUSION

The parametric WGEN and NWG are introduced in this chapter for stochastic downscaling. WGEN has been popularly employed in the assessment of hydrological and agricultural studies while NWG is a comparatively recent technique that is still developing for applying to future climate scenarios. There are a number of stochastic WGENs, and its selection should be based on the statistical characteristics of simulated series and an objective of a target study.

7 Weather-Type Downscaling

7.1 CLASSIFICATION OF WEATHER TYPES

7.1.1 Empirical Weather Type

Among other weather types, empirical weather typing requires extensive research and complete review of weather and atmospheric circulations. Here, the empirical weather typing by Lamb (1972) is presented, and its result is employed to generate daily rainfall sequences following the study of Wilby (1994).

Lamb (1972) subjectively classified weather types over the British Isles region (50°N–60°N and 10°W–2°E) into seven types: Anticyclonic (*A*), Cyclonic (*C*), Westerly (*W*), North-Westerly (*NW*), Northerly (*N*), Easterly (*E*), and Southerly (*S*), with the surface maps of pressure distribution, winds, and weather. This classification has been named as Lamb weather type (LWT). In Figure 7.1, the isobars of surface (left panels) and tropopause (right panels) for a few of the selected weather types are shown for the same days, as indicated by Lamb (1972). The top panels present the *A* type defined as mainly dry with light winds; the *C* type in the middle panels presents the condition of mainly wet or disturbed weather with variable wind directions and strengths; and the northerly type (*N*) on the bottom panels as cold disturbed weather at all seasons.

7.1.2 Objective Weather Type

Jones et al. (1993) suggested an objective classification with mean sea-level pressure (MSLP) in 16 grid points, as shown in Figure 7.2. The wind-flow characteristics are computed from each daily MSLP (denoted as ρ_i, $i = 1,\ldots,16$ with the unit of hPa and equivalent to 1.2 knots) as follows:

$$\omega = \frac{1}{2}(\rho_{12} + \rho_{13}) - \frac{1}{2}(\rho_4 + \rho_5) \tag{7.1}$$

$$\eta = 1.74\left[\frac{1}{4}(\rho_5 + 2\rho_9 + \rho_{13}) - \frac{1}{4}(\rho_4 + 2\rho_8 + \rho_{12})\right] \tag{7.2}$$

$$F = \sqrt{\omega^2 + \eta^2} \tag{7.3}$$

$$ZW = 1.07\left[\frac{1}{2}(\rho_{15} + \rho_{16}) - \frac{1}{2}(\rho_8 + \rho_9)\right] - 0.95\left[\frac{1}{2}(\rho_8 + \rho_9) - \frac{1}{2}(\rho_1 + \rho_2)\right] \tag{7.4}$$

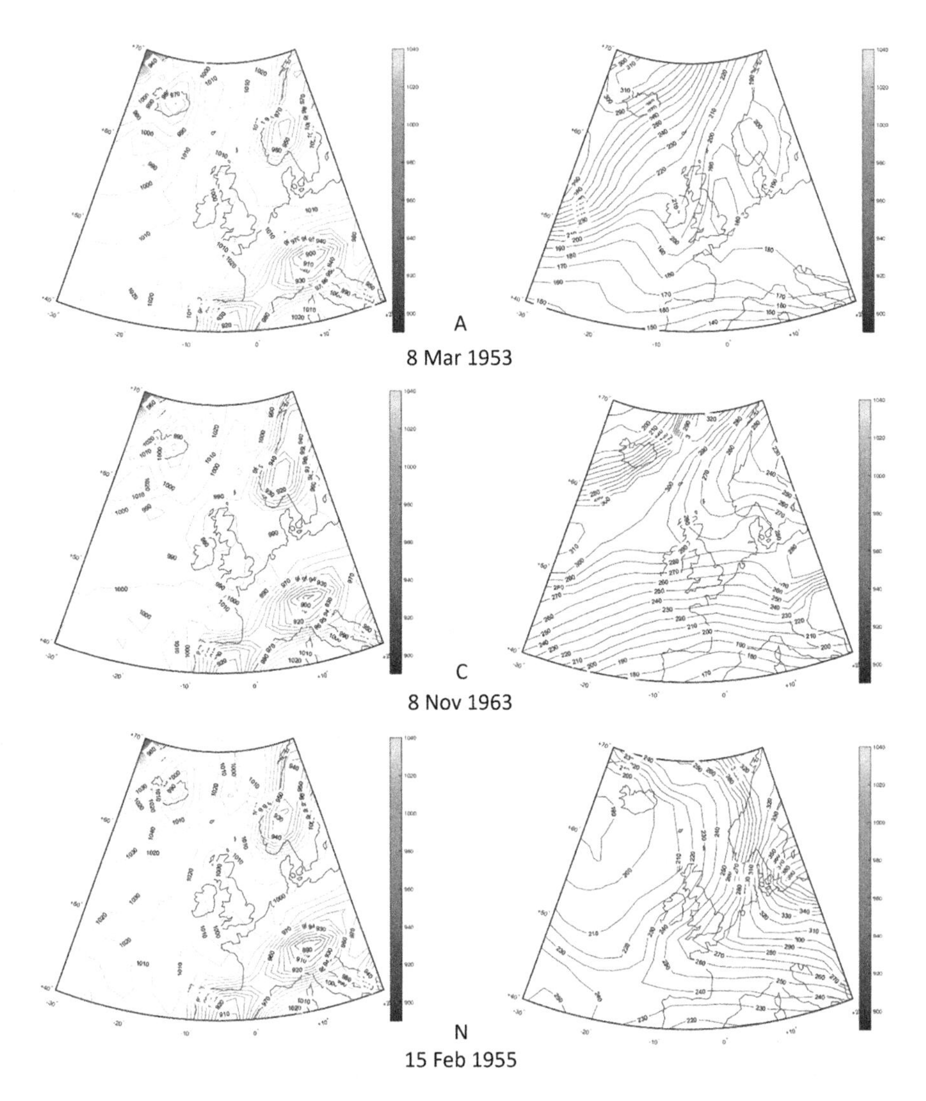

FIGURE 7.1 Isobars of surface (left panels) and tropopause (right panels) for the same days indicated in Lamb (1972) for different weather types, such as Anticyclonic (top panels), Cyclonic (middle), and Northerly (bottom). Note that pressure data was obtained from National Centers for Environmental Prediction (NCEP) reanalysis daily averages.

$$ZS = 1.52\left[\frac{1}{4}(\rho_6 + 2\rho_{10} + \rho_{14}) - \frac{1}{4}(\rho_5 + 2\rho_9 + \rho_{13}) - \frac{1}{4}(\rho_4 + 2\rho_8 + \rho_{12}) + \frac{1}{4}(\rho_3 + 2\rho_7 + \rho_{11})\right] \quad (7.5)$$

$$Z = ZW + ZS \quad (7.6)$$

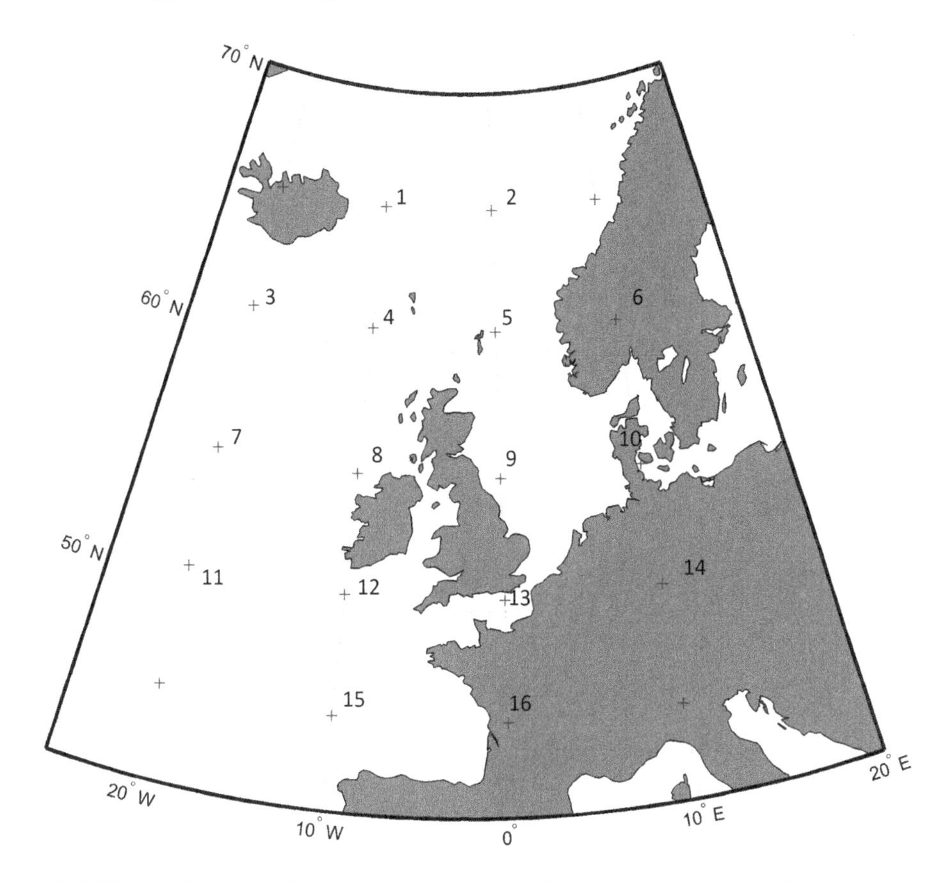

FIGURE 7.2 Location of the grid points over the British Isles to calculate the wind flow direction and vorticity terms.

Note that ω and η are westerly and southerly flows, respectively, while F is the resultant flow. ZW and ZS are westerly and southerly shear vorticities, respectively, while Z is the total shear vorticity.

The following rule is used by Jenkinson and Collison (1977) to define LWT:

1. The flow direction is

$$D = \tan^{-1}\left(\frac{\omega}{\eta}\right) \tag{7.7}$$

   Add 180° if ω is positive. The W weather type occurs when $247.5° < D < 292.5°$.
2. If $|Z|$ is less than F (i.e., $|Z| < F$), the flow is essentially straight and corresponds to an original Lamb type.
3. If $|Z|$ is greater than $2F$, then the pattern is strongly cyclonic ($Z > 0$) or anticyclonic ($Z < 0$).

4. If |Z| is lies between F and $2F$, then the flow is partly (anti)cyclonic, and this corresponds to one of the hybrid types.
5. If F is less than 6, and |Z| is less than 6, there is light indeterminate flow, corresponding to the unclassified type.

Example 7.1

Estimate ω and η as well as F and Z with the earlier equations for the dataset in Table 7.1. Also, define the weather type.

Solution:

a. The westerly and southerly winds (ω and η) are estimated with Eqs. (7.1) and (7.2):

$$\omega = \frac{1}{2}(1037.6 + 1026.4) - \frac{1}{2}(1032.1 + 1034.8) = -1.45$$

$$\eta = 1.74\left[\frac{1}{4}(1034.8 + 2\times 1030.4 + 1026.4)\right.$$

$$\left. - \frac{1}{4}(1032.1 + 2\times 1031.5 + 1037.6)\right] = -4.655$$

b. The resultant flow and total shear vorticity (F and Z) are estimated with Eqs. (7.3) and (7.6) as $F = 4.875$ hPa and $Z = -91.479$ hPa.
c. The flow direction with Eq. (7.7) is

$$D = \tan^{-1}\left(\frac{-1.45}{-4.655}\right) = 0.302$$

TABLE 7.1
MSLP on March 8, 1953 for the Same Location as in Figure 7.2

	Num	20°W	10°W	0°	10°E
Location Number	65°N		1	2	
	60°N	3	4	5	6
	55°N	7	8	9	10
	50°N	11	12	13	14
	45°N		15	16	
	MSLP (hPa)	**20°W**	**10°W**	**0°**	**10°E**
Current Obs. (March 8, 1953)	65°N		1015.7	1005	
	60°N	1028.7	1032.1	1034.8	952.4
	55°N	1032.8	1031.5	1030.4	1028.2
	50°N	1028.9	1037.6	1026.4	989.5
	45°N		1022.2	999.9	

The numbers in the upper part of the table indicate the locations of Figure 7.2, and the corresponding MSLPs are presented in the lower part of the table. The data is obtained from NCEP reanalysis at www.esrl.noaa.gov/psd/data/gridded/data.ncep.reanalysis.html.

Because the estimated value is radian, it should be conversed to degree as $D(\text{deg}) = 0.302 \times \frac{180}{\pi} = 197.3°$

d. Since $|Z|$ (91.479) is larger than $2F$ and $Z < 0$, the weather type is categorized as anticyclonic, (i.e., Type A). See the top panel of Figure 7.1 for more detailed isobars.

7.2 GENERATION OF DAILY RAINFALL SEQUENCES

Employing the LWT catalog (Lamb, 1972), transition matrices and precipitation statistics can be extracted. Tables 7.2 and 7.3 show the assumed example of the results by Wilby (1994) for the site of Kempsford in the Cotswolds Region of England, respectively, and the statistics leads from the weather-type data using Section 7.1 and Example 7.1 similar to Lamb (1972), but including O weather type that is not categorized as three types (A, E, and C) like Wilby (1994). Once the weather type is defined, it is straightforward to estimate those statistics. Daily rainfall series is generated with the matrices and statistics as follows:

TABLE 7.2
Example of Assumed Transition Matrices of LWT Similar to the One of Wilby (1994)

Weather Type	*A*	*E*	*C*	*O*
A	0.50	0.05	0.02	0.43
E	0.10	0.45	0.05	0.40
C	0.15	0.05	0.05	0.30
O	0.20	0.05	0.20	0.55

Here, only three types are considered, while Wilby (1994) took eight types into account. Note that O presents the type that is not categorized into the other three types. For example, the second column and the second row value of 0.05 refers to the probability that occurs with the E type with the previous condition being the A type.

TABLE 7.3
Assumed Precipitation Statistics Conditioned on the Defined LWT of Table 7.2

Weather Type	P (LWT)	P_{wet}	μ_{LWT} (mm)
A	0.35	0.13	1.8
E	0.12	0.60	3.5
C	0.08	0.80	5.5
O	0.45	0.45	4.5

The second column presents the occurrence probability of each LWT, the third column is the probability for precipitation occurrence, and the fourth column is the daily average precipitation.

1. At first, the weather type is simulated with either the transition matrices of Markov chain (see Section 6.1.2) in Table 7.2 or the occurrence probability of LWT in the second column of Table 7.3. The latter is employed when the weather type of the preceding day is not defined (i.e., the first day of simulation). Otherwise, the former is employed. For example, the weather type of the first day is generated from the cumulative distribution of the occurrence probability of the LWT type with the order of [*A*, *E*, *C*, *O*], as in the order of Table 7.3: $P_C^{TM} = [0.35, 0.47, 0.55, 1.000]$

   Here, the *O* type refers to atmospheric patterns that are not defined with the earlier seven types.
2. A uniform number ($u \sim U[0,1]$) is simulated as in Section 2.3.1 and compared with P_C^{TM}. According to the generated random number, a weather type is assigned as {[0 0.35) *A*; [0.35 0.470) *E*;...; [0.470 0.55] *C*; [0.55 1.00] *O*}. Note that [*a b*) indicates the range that "*a*" is inside the range but *b* is not. Assume that $u = 0.441$ is generated, its value is in the range of [0.35 0.470), and then the *E* type is assigned. The following type is generated from the cumulative probabilities in the third row of Table 7.2 with the order of [*A*, *E*, *C*, *O*] since its previous condition is the *E* type:

$$P_C^{TM} = [0.10, 0.55, 0.60, 1.00]$$

   Assume that $u = 0.511$ is simulated and then, it is in the range of [0.10 0.55) and the type *E* is assigned again.
3. From the assigned weather type (here *E*), the occurrence of the precipitation is simulated with the uniform number and the P_{wet} (here 0.60) in the third column of Table 7.3. Say, $u = 0.322$ and since $P_{wet} < u$ ($0.60 < 0.322$), the current day is wet. If the current day is wet, then the following is performed for the rainfall amount. Otherwise, the rainfall amount is assigned as zero.
4. To simulate the rainfall amount, gamma or exponential distributions have been employed. Here, exponential distribution from Wilby (1994) is used as

$$R = -\mu_{LWT} \ln(r)(1+u) \tag{7.8}$$

   where u and r are the uniform random numbers and μ_{LWT} is the mean daily rainfall amount in the fourth column of Table 7.3. Assume that r and u are simulated as 0.512 and 0.771 from the uniform distribution. The rainfall amount is

$$R = -\mu_{LWT} \ln(r)(1+u) = -3.5\ln(0.512)(1+0.771) = 4.15\,\text{mm}$$

5. Repeat steps (2)–(4) until the required simulation is done.

7.3 FUTURE CLIMATE WITH WEATHER-TYPE DOWNSCALING

Future revolution of daily precipitation is simulated by conditioning the weather type obtained from global climate model (GCM) outputs. Weather conditions can be defined from GCM outputs (e.g., MSLP) to classify weather types with the following simple steps.

1. Obtain future atmospheric circulation variables (e.g., MSLP) from GCM outputs.
2. Classify the future daily weather types with the atmospheric weather variables from GCM outputs. Note that the criteria of weather types must be predefined from the observed atmospheric weather variables (e.g., MSLP) in the same way of Section 7.1.
3. According to the defined weather types, daily precipitation of future period is simulated as in Section 7.2.

7.4 SUMMARY AND CONCLUSION

The weather-type downscaling is briefly explained in this chapter. One of the major advantages of the weather-type downscaling is to relate physical weather conditions in statistical downscaling by defining weather types. To properly connect weather types in statistical downscaling, a classification method for weather type is critical. Here, a rather simple empirical approach is introduced. Further complex classification methods can be applied such as k-means (MacQueen, 1967), machine learning (Leloup et al., 2008), and Gaussian mixture distributions (Rust et al., 2010).

8 Temporal Downscaling

8.1 BACKGROUND

8.1.1 K-NEAREST NEIGHBOR RESAMPLING

The *K*-nearest neighbor resampling (KNNR) method was developed by Lall and Sharma (1996) for generating hydrologic time series. The background of this approach lies in a *K*-nearest neighbor density estimator that employs the Euclidean or Mahalanobis distance to the *K*th nearest data point and its volume containing *K*-data points. The KNNR resamples a value from the observed data according to the closeness of the distance between the provided values and observations of the predictor variables, denoted as feature vector.

Consider the multivariate predictor variable vector $\mathbf{x}_i = \left[x_i^{(1)}, x_i^{(2)}, \ldots, x_i^{(s)}\right] = \left[x_i^{(s)}\right]_{s \in \{1,S\}}$, where S is the number of participating variables; $i = 1, \ldots, n$, where n is the observed record length. The corresponding predictor variable is y_i; $i = 1, \ldots, n$. Both $\mathbf{x}_i$ and y_i, $i = 1, \ldots, n$, are known for the observation period. The objective is to simulate the predictand variable y at time t, denoted as Y_t. Provided that the predictor variables at time t (i.e., $\mathbf{x}_t$) and the number of neighbors (k) are known, the procedure is described as follows:

1. Estimate the distances between the conditional feature vector (or a set of predictor variables at time t) and the corresponding observation vector $\mathbf{x}_i$. Here, the distances are measured for $i = 1, \ldots, n$ as

$$D_i = (\mathbf{X}_t - \mathbf{x}_i)^T \mathbf{C}^{-1} (\mathbf{X}_t - \mathbf{x}_i)^T \tag{8.1}$$

   where $\mathbf{C}^{-1}$ is the inverse of the covariance matrix of $\mathbf{x}$ ($\mathbf{C}$). This distance measurement is called the Mahalanobis distance. Note that if the covariances between the weather variables are neglected (i.e., $\mathrm{Cov}(x^{(p)}, x^{(q)}) = 0$, when $p \neq q$), $\mathbf{C}$ becomes a diagonal matrix and its diagonal elements are the variances of each variable (i.e., $\left[\sigma_{x^s}^2\right]_{s \in \{1,S\}}$). In case that each variable has the same variance and set $\left[\sigma_{x^s}^2\right]_{s \in \{1,S\}} = 1$, this measurement is the Euclidean distance as $D_i = (\mathbf{X}_t - \mathbf{x}_i)(\mathbf{X}_t - \mathbf{x}_i)^T$.
2. Arrange the estimated distances from step (1) in ascending order, select the first k smallest distances and store the time indices of the smallest k distances.
3. Randomly select one of the stored k time indices with the weighting probability given by

$$w_m = \frac{1/m}{\sum_{j=1}^{k} 1/j}, \qquad m = 1,\ldots,k \tag{8.2}$$

4. Assume that the time index p is selected and assign the corresponding predictor variable Y_t as the selected time index. Then, $Y_t = y_p$

The number of nearest neighbors, k, can be estimated with different approaches (Lee and Ouarda, 2011). One of the most common approaches in literature is the heuristic selection as $k = \sqrt{n}$ (Lall and Sharma 1996). The role of the number of nearest neighbors is to limit the number of selected neighbors by force. However, its sensitivity to the choice of k is small.

Example 8.1

Simulate the January precipitation (unit: mm) of Cedars, Quebec, as in Table 5.7, with KNNR. The explanatory variable is the wind direction obtained from the National Centers for Environmental Prediction/National Center for Atmospheric Research (NCEP/NCAR)analysis data like Table 5.4. The total dataset is shown in Table 8.1.

Solution:

The number of nearest neighbors is estimated with the heuristic approach as $k = \sqrt{n} = \sqrt{26} = 5.099 \overset{ceil}{\approx} 6$. Here, $\overset{ceil}{\approx}$ represents to map the given value to the least integer that is greater than or equal to the given value. Assume the given feature vector at the target time t, i.e., $x_t = 90°$. The following procedure is used:

1. Estimate the distances between the conditional feature vector (here, $x_t = 90°$) and the corresponding observations as in the third column of Table 8.1. The distances are measured for $i = 1, \ldots, n$ as $D_i = (x_t - x_i)^2$. The distance equation is simplified because only one predictor variable is employed. For example, $D_i = (90–92.39)^2 = 5.73$. The estimated differences are presented in the fifth column of Table 8.1.
2. Sort the estimated distances from step (1) in ascending order. Its sorted order and the corresponding differences are shown in the sixth and seventh columns of Table 8.1, respectively.
3. Select the first k (here, 6) smallest distances and store the time indices of the smallest k distances (here, 2, 4, 1, 20, 13, 19).
4. Randomly select one of the stored k time indices with the weighting probability estimated with Eq. (8.2) as $\mathbf{w} = [0.4082, 0.2041, 0.1361, 0.1020, 0.0816, 0.068]$. Assume six is chosen, and the sixth time index is $i = 19$ as in step (2).
5. Assign the corresponding predictor variable Y_t as the selected time index. Then, $Y_t = y_{(6)} = y_{19} = 58.92$.

   The simulated data with the KNNR is shown in Table 8.1 employing the observed predictor value as the provided current predictor values (i.e., $x_t = x_i$). Also, Figure 8.1 illustrates that the simulated precipitation with KNNR presents the characteristics of observed data. At the top panel of Figure 8.1, the simulated data is scattered with a pattern similar to that of the observed data. The observed and simulated precipitation are well aligned, as shown in the bottom panel of Figure 8.1.

TABLE 8.1
Observed Dataset for the KNNR Example (Left Side) and the KNNR Procedure Data (Right Side) as well as the Simulated Dataset in the Last Column

		Observed		Simulation Procedure			New Simulated Dataset
Order	Year	Wind Direction (°)	Precipitation (°C)	Difference	Sorted Order	Sorted Difference	Sim. Precipitation
1	1961	92.39	58.69	5.73	2	1.29	58.92
2	1962	91.14	42.11	1.29	4	2.95	48.73
3	1963	140.16	57.13	2515.88	1	5.73	57.13
4	1964	88.28	69.47	2.95	20	12.18	69.47
5	1965	129.69	53.64	1575.47	13	32.56	50.74
6	1967	123.43	66.65	1117.67	19	36.77	48.65
7	1968	105.29	58.51	233.73	21	64.43	58.51
8	1969	176.97	72.80	7563.99	11	65.09	79.48
9	1970	77.52	50.95	155.79	22	67.13	50.95
10	1971	130.49	48.65	1639.31	12	67.91	66.65
11	1972	98.07	49.49	65.09	14	126.55	49.49
12	1973	98.24	62.19	67.91	9	155.79	58.49
13	1974	84.29	50.01	32.56	24	172.91	50.01
14	1975	101.25	64.39	126.55	7	233.73	82.55
15	1976	106.86	52.24	284.37	15	284.37	58.51
16	1977	140.21	50.74	2520.84	25	369.18	57.13
17	1978	159.61	69.29	4844.89	23	408.31	69.29
18	1979	167.52	79.48	6009.77	26	617.14	69.29
19	1980	96.06	58.92	36.77	6	1117.67	58.92
20	1981	86.51	48.73	12.18	5	1575.47	58.69
21	1984	98.03	62.48	64.43	10	1639.31	62.48
22	1985	98.19	58.49	67.13	3	2515.88	62.19
23	1986	110.21	63.05	408.31	16	2520.84	63.05
24	1987	103.15	82.55	172.91	17	4844.89	62.19
25	1989	109.21	60.60	369.18	18	6009.77	60.6
26	1990	114.84	69.85	617.14	8	7563.99	69.85

8.2 DAILY TO HOURLY DOWNSCALING

While time-series data are essential for assessing the hydrological effects of climate change on medium- or small-sized watersheds, the time series of hydrometeorological variables of interest, such as precipitation (or rainfall), are not always available at the desired time. Therefore, hydrometeorological data with time scales on the order of 1 h or less are urgently required. To obtain precipitation time series with finer temporal resolution, the creation of hourly time series from daily time series using disaggregation rules, such as preserving the diurnal cycle and the additive condition, is recommended (Jones and Harpham, 2009).

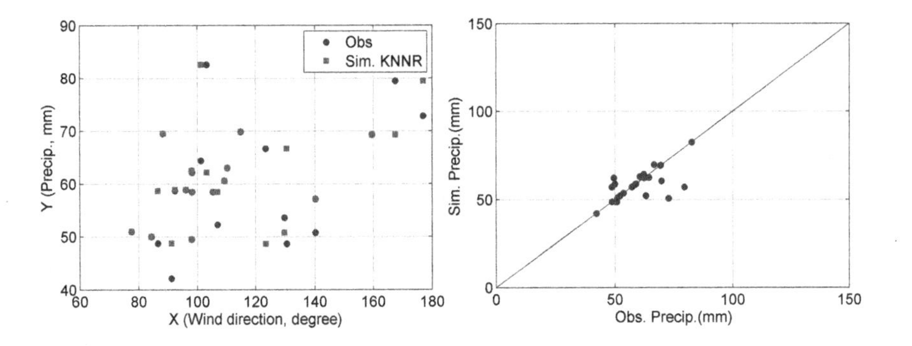

FIGURE 8.1 Example of *K*-nearest neighbor resampling for the January precipitation at Cedars, Quebec, with the predictor variable of the wind direction for the NCEP/NCAR analysis data. (1) Left panel: the scatterplot of the predictor variable (wind direction, degree) and the predictand variable (January precipitation, mm) for observed data (blue circles) and simulated data (red squares) and (2) right panel: the scatterplot of observed predictand data (x-axis) versus simulated data (y-axis).

Several approaches for statistical temporal downscaling of precipitation time series have been suggested in literature. Among others, Lee and Jeong (2014) presented a theoretical framework for a nonparametric temporal downscaling model that considers the diurnal cycle and the specific details of key hourly statistics using KNNR and the Genetic Algorithm (GA) mixing process. The method is considerably suited for exploring future climate scenarios and assessing the impacts of climate change for flood control management. In addition, the method can be easily adapted to any hydrometeorological variable, which is especially beneficial for the reproduction of distinct precipitation subdaily patterns. This downscaling procedure by Lee and Jeong (2014) is presented in the following.

Consider daily, y_i, and hourly observations $\mathbf{x}_i = \left[x_{i,1}, x_{i,2}, \ldots, x_{i,24}\right] = \left[x_{i,h}\right]_{h\in\{1,24\}}$ and $i = 1, \ldots, n$, where n is the record length and h indicates the hth hour. In addition, y_t is the target daily precipitation value for time t for $t = 1, \ldots, T$, where T is the length of the target daily precipitation. Note that these target daily precipitation values are already known. The objective is to downscale the daily time series y_t to the hourly time series $\mathbf{x}_t = \left[x_{t,1}, x_{t,2}, \ldots, x_{t,24}\right]$. Provided that the number of nearest neighbors, k, is known, the temporal downscaling procedure is as follows.

1. Estimate the distance between the target daily precipitation y_t and the observed daily precipitation $y_i = \sum_{h=1}^{24} x_{i,h}$. Here, the distances are measured for $i = 1, \ldots, n$ as

$$D_i = (y_t - y_i)^2 \tag{8.3}$$

2. Arrange the estimated distances from step (1) in ascending order, select the first k distances (i.e., the smallest k values), and reserve the time indices of the smallest k distances.

3. Randomly select one of the stored k time indices with the weighting probability given as

$$w_m = \frac{1/m}{\sum_{j=1}^{k} 1/j}, \qquad m = 1,\ldots,k \tag{8.4}$$

4. Assign the hourly values of the selected time index from step (3) as $\mathbf{x}_p = [x_{p,h}]_{h\in\{1,24\}}$. Here, it is assumed that the selected time index is p.
5. Execute the following steps for GA mixing:
   (5.1) Reproduction: Select one additional time index using steps (1)–(4) and denote this index as p^*. Obtain the corresponding hourly precipitation values, $\mathbf{x}_{p^*} = [x_{p^*,h}]_{h\in\{1,24\}}$. The subsequent two GA operators employ the two selected vectors, $\mathbf{x}_p$ and $\mathbf{x}_{p^*}$.
   (5.2) Crossover: Replace each element $x_{p,h}$ with $x_{p^*,h}$ at probability P_c, i.e.,

$$X_{t,h}^* = \begin{cases} x_{p^*,h} & \text{if } \varepsilon < P_c \\ x_{p,h} & \text{otherwise} \end{cases}, \tag{8.5}$$

   where ε is a uniformly distributed random number between 0 and 1.
   (5.3) Mutation: Replace each element (i.e., each hour, $h = 1, \ldots, 24$) with one selected from all the observations of this element for $j = 1, \ldots, n$ (hourly precipitation data) with probability P_m, i.e.,

$$X_{t,h}^* = \begin{cases} x_{\xi,h} & \text{if } \varepsilon < P_c \\ x_{p,h} & \text{otherwise} \end{cases}, \tag{8.6}$$

   where $x_{\xi,h}$ is selected from $[x_{i,h}]_{i\in\{1,n\}}$ with equal probability for $i = 1, \ldots, n$.
6. Adjust the GA mixed hourly values to preserve the additive condition as follows:

$$X_{t,h} = \frac{X_{t,h}^*}{\sum_{j=1}^{24} X_{t,j}^*} Y_t \tag{8.7}$$

   where $X_{t,h}$ is the downscaled hourly time series including the GA mixture process, $h = 1, \ldots, 24$. Note that the additive condition implies the summation of simulated hourly precipitation for 24 h must be the same as daily precipitation.
7. Repeat steps (1)–(6) until the required data are generated.

The roles of crossover probability P_c and mutation probability P_m were studied by Lee et al. (2010). These two probabilities can be employed as tuning parameters to improve the reproduction of historical statistics for the generated hourly time series. In practice, $P_c = 0.1$ and $P_m = 0.01$ are employed without significant sensitivity (Lee et al., 2010). The selection of the number of nearest neighbor (k) has been studied (Lall and Sharma, 1996; Lee and Ouarda, 2011). The most common and simplest selection method was applied by setting $k = \sqrt{n}$.

Example 8.2

Downscale the target daily precipitation (assuming the target daily precipitation is 50 mm, $x_t = 50$) to hourly precipitation with the observed dataset in Table 8.2.

Solution:

The partial observed daily and hourly precipitation of 1987 is displayed in Table 8.2 for the Jinju weather station in South Korea. Note that the dates with zero daily

TABLE 8.2
Daily and Hourly Observed Precipitation (mm) at Jinju, South Korea, in 1987

			Hourly											
Month	**Day**	**Daily**	**1**	**2**	**3**	**4**	**5**	**6**	**7**	**8**	**9**	**10**	**11**	**12**
7	6	15.4	2.1	3.3	7.6	2.4	0	0	0	0	0	0	0	0
7	11	27	0	0	0	0	0	0	0	0	0	5	3.3	2.4
7	12	40.8	0	0	0	0	0	0	0	0	0	0	0	0
7	13	58.5	3.2	2.6	2.6	7.6	9.5	5.8	3.9	4.2	1.9	2.5	1.5	1.9
7	14	13.5	0.2	0.1	0.2	0	0	0.1	0.1	0	0	0	0	0
7	15	188.3	0.2	1.2	1.4	2.9	1.9	2.7	0.3	0	0	0	0	0
7	16	0.5	0.5	0	0	0	0	0	0	0	0	0	0	0
7	18	2.3	0	0	0	0	0	0	0	0	0	0	0	0
7	21	0.1	0	0	0	0	0.1	0	0	0	0	0	0	0
7	22	14.5	0	0	0.1	0	0.1	0.2	0.1	0	0	0	0	0
7	23	35.8	0.1	1.8	0.1	1.5	**3.7**	0.7	13	1.1	2.6	0.8	2.8	0.2
7	24	9.6	0.5	0.1	0.5	0.5	0.1	0.1	0	0.1	0.6	1.8	0.3	0.2
7	25	19.7	0	0	0	0	0	0.1	0.1	0.1	0.3	2	2.9	5.8
7	26	8.2	0.8	0.1	0.1	1.3	0	0	0.2	0.1	0	3.6	1.4	0
7	27	7.1	0	0	0	0	0	0	0	6.5	0.6	0	0	0
8	2	5.9	0	0	0	0	0	0	0	0.4	1.3	0.3	0	0
8	3	6.2	0	0	0	0	0	0	0.1	0	0	0	0	0
8	4	6.3	0	0	0	0	0	0	0	0	0	0	0	0
8	5	16.3	16	0.1	0.5	0	0	0	0	0	0	0	0	0
8	7	14.9	0	0	0	0	0	0	0	0	0	0	0	0
8	10	5.4	0	0	0	0	0.5	3.4	1.4	0.1	0	0	0	0
8	11	16.2	0	5.5	0	0	0	0	0	0	0.6	0.8	1.2	1.5
8	14	4.6	0	0	0	0	0	0	0	0	0	3.6	0	0.1
8	15	44.6	0	0.1	0.5	0.7	0	1.1	1	2	9.6	24	2.1	1.5
8	16	33.1	0	0	0.3	1.5	0	1.2	0	0	0	0	0	0
8	17	17.9	0.3	0	0	0	2.3	4.4	4.1	3.1	1.9	1.3	0.4	0.1
8	21	36.1	0	0	0.3	1.4	2.8	7.1	9	7.1	7.3	1.1	0	0
8	22	1.3	0	0	0.1	0.1	0.1	0.2	0.7	0.1	0	0	0	0

(Continued)

TABLE 8.2 (*Continued*)
Daily and Hourly Observed Precipitation (mm) at Jinju, South Korea, in 1987

			Hourly											
Month	Day	Daily	13	14	15	16	17	18	19	20	21	22	23	24
7	6	15.4	0	0	0	0	0	0	0	0	0	0	0	0
7	11	27	0.7	1	3.6	3.4	1.6	1	3.9	1	0.1	0	0	0
7	12	40.8	0	0.7	0	1.2	2.3	0.3	5.6	5	3.3	4.6	10.5	7.3
7	13	58.5	0.6	1.2	0.1	0.8	0	0	3.7	3.5	0.9	0.3	0.1	0.1
7	14	13.5	0	0	0	2.6	2.2	2.5	1.8	1.8	0.9	0.8	0.2	0
7	15	188.3	0	0.2	0.6	11.2	15.4	24.1	15.3	15.4	21.8	35.8	31.2	6.7
7	16	0.5	0	0	0	0	0	0	0	0	0	0	0	0
7	18	2.3	0	0	0	0	0	0	0.4	1	0.5	0.1	0.2	0.1
7	21	0.1	0	0	0	0	0	0	0	0	0	0	0	0
7	22	14.5	0	0	0	0	0	0	0	10	3.2	0.5	0.2	0.1
7	23	35.8	0	0	0	0.4	0.1	0.6	0.4	0.1	0.4	1.8	2.4	1.6
7	24	9.6	0.5	1.7	0.5	0.7	0.1	0.3	0.9	0.1	0	0	0	0
7	25	19.7	2.6	1.9	1.1	1	1.4	0	0.1	0	0.2	0	0	0.1
7	26	8.2	0	0.3	0.3	0	0	0	0	0	0	0	0	0
7	27	7.1	0	0	0	0	0	0	0	0	0	0	0	0
8	2	5.9	0.6	3.3	0	0	0	0	0	0	0	0	0	0
8	3	6.2	0	0	0.1	0	0	0	0	2.5	0.8	0	2.5	0.2
8	4	6.3	0	0	1.4	0	0	0	0	0	0	0	4	0.9
8	5	16.3	0	0	0	0	0	0	0	0	0	0	0	0
8	7	14.9	2	2.9	0.5	0.6	1.1	2.8	5	0	0	0	0	0
8	10	5.4	0	0	0	0	0	0	0	0	0	0	0	0
8	11	16.2	2	1.5	1.3	1.6	0.2	0	0	0	0	0	0	0
8	14	4.6	0	0.1	0	0	0.2	0.4	0	0	0	0	0.2	0
8	15	44.6	2.2	0.2	0	0	0	0	0	0	0	0	0	0
8	16	33.1	0	0	0	12	11.8	0.6	0	0	2.4	0.7	1.3	1.3
8	17	17.9	0	0	0	0	0	0	0	0	0	0	0	0
8	21	36.1	0	0	0	0	0	0	0	0	0	0	0	0
8	22	1.3	0	0	0	0	0	0	0	0	0	0	0	0

rainfall are omitted from the dataset. The first and second columns present the date of record and daily precipitation is in the third column followed by 24 hourly precipitation for each date.

The heuristic estimation of k is $k = \sqrt{n} = \sqrt{28} = 5.2915 \overset{ceil}{\approx} 6$.

1. Estimate the distances between target daily precipitation y_t and observed daily precipitation. For example, $D_1 = (50-15.4)^2 = 1197.16$ for the first observed day in the third row of Table 8.2. All the distances are shown in the third column of Table 8.3.
2. Arrange the estimated distances in ascending order as in the fourth and fifth columns of Table 8.3 and reserve the time indices of the smallest k distances (here, $k = 6$ and the time indices are 24, 4, 3, 27, 11, 25).

TABLE 8.3
Example of Nonparametric Downscaling from Daily to Hourly

Order	Daily	Dist	Sort Order	Sort Dist
1	15.4	1197.16	24	29.16
2	27	529	4	72.25
3	40.8	84.64	3	84.64
4	58.5	72.25	27	193.21
5	13.5	1332.25	11	201.64
6	188.3	19126.89	25	285.61
7	0.5	2450.25	2	529
8	2.3	2275.29	13	918.09
9	0.1	2490.01	26	1030.41
10	14.5	1260.25	19	1135.69
11	**35.8**	201.64	22	1142.44
12	9.6	1632.16	1	1197.16
13	19.7	918.09	20	1232.01
14	8.2	1747.24	10	1260.25
15	7.1	1840.41	5	1332.25
16	5.9	1944.81	12	1632.16
17	6.2	1918.44	14	1747.24
18	6.3	1909.69	15	1840.41
19	16.3	1135.69	18	1909.69
20	14.9	1232.01	17	1918.44
21	5.4	1989.16	16	1944.81
22	16.2	1142.44	21	1989.16
23	4.6	2061.16	23	2061.16
24	**44.6**	29.16	8	2275.29
25	**33.1**	285.61	28	2371.69
26	17.9	1030.41	7	2450.25
27	**36.1**	193.21	9	2490.01
28	1.3	2371.69	6	19126.89

Note that the target daily precipitation value is 50 mm and $k = 6$. Daily and hourly observed precipitation (mm) in Jinju, South Korea, in 1987 as shown in Table 8.2.

3. Randomly select one of the stored k time indices with the weighting probability in Eq. (8.4). Assume that the fourth distance and its corresponding index (i.e., (2) = 4, the second smallest distance ($D_{(2)} = D_4 = 72.25$) as in the fourth column of Table 8.3 is selected, whose order is 4 shown in the first column).
4. Assign hourly values to the selected time index from $\mathbf{x}_4 = [x_{4,h}]_{h \in \{1,24\}} = [3.2, 2.6, 2.6, \ldots, 0.1]$.
5. Execute the following steps for GA mixing:
   (5.1) Reproduction: Select one additional time index and assume $p^* = (4) = 27$. Obtain the corresponding hourly precipitation values, $\mathbf{x}_{27} = [x_{27,h}]_{h \in \{1,24\}} = [0, 0, 0.3, 1.4, \ldots, 0]$. The subsequent two GA operators employ the two selected vectors, $\mathbf{x}_p = \mathbf{x}_4$ and $\mathbf{x}_{p^*} = \mathbf{x}_{27}$,

TABLE 8.4
Two Selected Hourly Precipitation (mm) Datasets in the Second and Third Columns with KNNR and Their GA-Processed Dataset as Crossover (Fourth) and Mutation (Fifth) as well as the Final-Adjusted Hourly Precipitation (mm) to Meet the Additive Condition of 50 mm

Hours	Sel-1 (4)	Sel-2 (27)	Crossover	Mutate	Adjusted
1	3.2	0	3.2	3.2	2.909
2	2.6	0	2.6	2.6	2.364
3	2.6	0.3	2.6	3.7	3.364
4	7.6	1.4	1.4	1.4	1.273
5	9.5	2.8	9.5	9.5	8.636
6	5.8	7.1	5.8	5.8	5.273
7	3.9	9	9	9	8.182
8	4.2	7.1	4.2	4.2	3.818
9	1.9	7.3	1.9	1.9	1.727
10	2.5	1.1	2.5	2.5	2.273
11	1.5	0	1.5	1.5	1.364
12	1.9	0	1.9	1.9	1.727
13	0.6	0	0.6	0.6	0.545
14	1.2	0	1.2	1.2	1.091
15	0.1	0	0.1	0.1	0.091
16	0.8	0	0.8	0.8	0.727
17	0	0	0	0	0.000
18	0	0	0	0	0.000
19	3.7	0	3.7	3.7	3.364
20	3.5	0	**0**	0	0.000
21	0.9	0	0.9	0.9	0.818
22	0.3	0	0.3	0.3	0.273
23	0.1	0	0.1	0.1	0.091
24	0.1	0	0.1	0.1	0.091
			Sum	55	50

presented in the second and third columns of Table 8.4. In Figure 8.2, the two selected hourly time series are illustrated with thin dash-dotted and dashed lines with circles. The first set (Set-1) of hourly precipitation data shows that the major rainfall event occurs in the first 10 hours followed by the smaller rainfall event for the next 19–21 hours but the rainy hours are sustained for the whole day. The second set (Set-2) presents that the rain event occurs for the first 10 hours and the other hours show no rain.

(5.2) Crossover: Replace each hour ($h = 1, \ldots, 24$) $x_{4,h}$ with $x_{27,h}$ at probability $P_c = 0.1$ as in Eq. (8.5). For example, at $h = 1$, if $\varepsilon = 0.531 > P_c = 0.1$, then $X_{t,1}^* = x_{p,1} = x_{4,1} = 3.2$. At $h = 4$, $\varepsilon = 0.055$ and $X_{t,4}^* = x_{p^*,4} = x_{24,4} = 1.4$. See Table 8.4 and Figure 8.2 for detail.

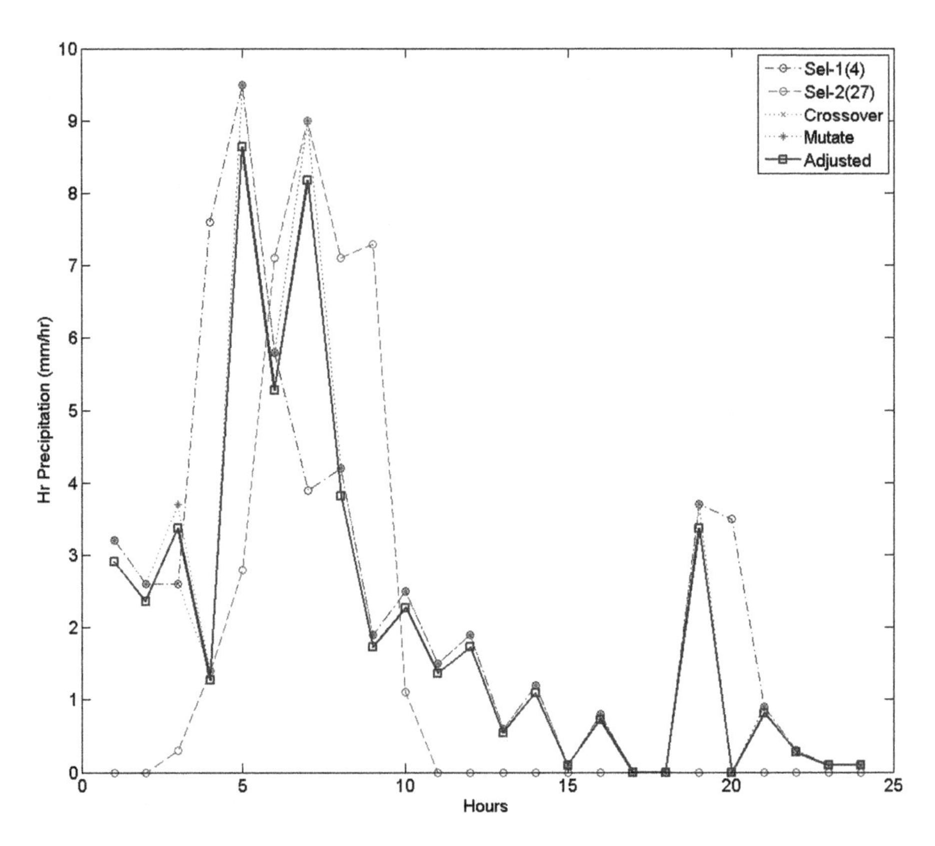

FIGURE 8.2 Example of the downscaling procedure of daily precipitation to hourly precipitation for the weather station of Jinju, South Korea, targeting 50 mm of daily precipitation.

(5.3) Mutation: Replace each hour (i.e., each hour, $h = 1, \ldots, 24$) with one selected from all the observations of this element as in Eq. (8.6) with the crossover hourly data shown in the fourth column of Table 8.4 and the dotted line with x marker in Figure 8.2. Note that the number of nonzero hourly precipitation is $n_h = 223$ among 672 values (i.e., 28×24) and the rest of the values are zero. For example, $\varepsilon = 0.008 < P_m(=0.01)$ for $h = 3$ (see the fifth column of Table 8.4), then select a value from all the observed values for $j = 1, \ldots, n_h$. Assume that $X^*_{t,h} = 3.7$ is selected in the fifth hour of 7/23, as shown in Table 8.2. The other hours are supposedly not selected, because the generated random numbers are $\varepsilon > P_m(= 0.01)$. See the dotted line with + marker in Figure 8.2.

6. Adjust the GA mixed hourly values in the fifth column of Table 8.4

$$X_{t,h} = \frac{X^*_{t,h}}{\sum_{j=1}^{24} X^*_{t,j}} Y_t = \frac{3.2}{55} 50 = 2.91$$

This adjustment must be made to preserve the additive condition of the daily value (i.e., 50 mm) with the summation of the hourly data. The adjusted downscaled hourly precipitation is presented in the fifth column of Table 8.4 as well as with the solid line with squares in Figure 8.2. The hourly pattern of the new downscaled precipitation presents two peaks in the fifth and seventh hours, followed by a smaller peak in the last hours.

8.3 SUMMARY AND CONCLUSION

The temporal downscaling of the precipitation output of climate models is critical in hydrological assessment of climate change, especially for small or urban watersheds. Nonparametric-based modeling of temporal downscaling successfully produces hourly precipitation conditioned on daily precipitation values. Temporal downscaling of monthly to daily can be a useful tool, but its applications are getting limited since most of the current climate models produce daily outputs already.

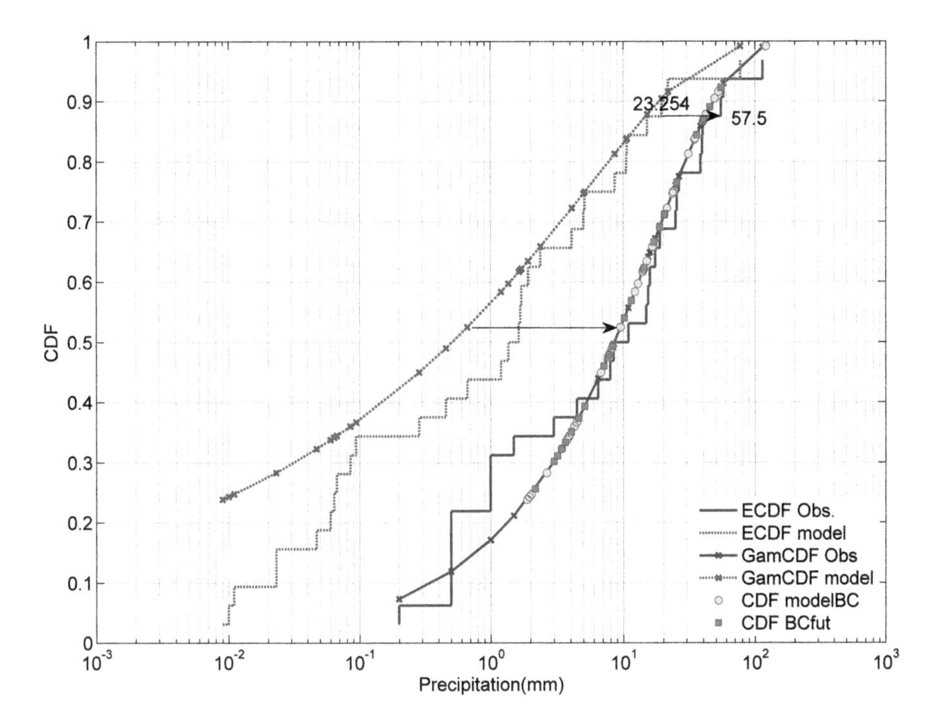

FIGURE 4.1 QM with gamma and empirical distributions (see Table 4.3). The Gamma CDFs for observed and GCM data are presented with smooth red solid and blue dotted lines with cross makers. The specific values for the bias-corrected data for the observed period (CDF model BC) and future period (CDF BCfut) are shown with circle and square markers. In addition, the ECDFs of the observed and GCM model data are shown with the stairs as red solid and blue dotted lines, respectively.

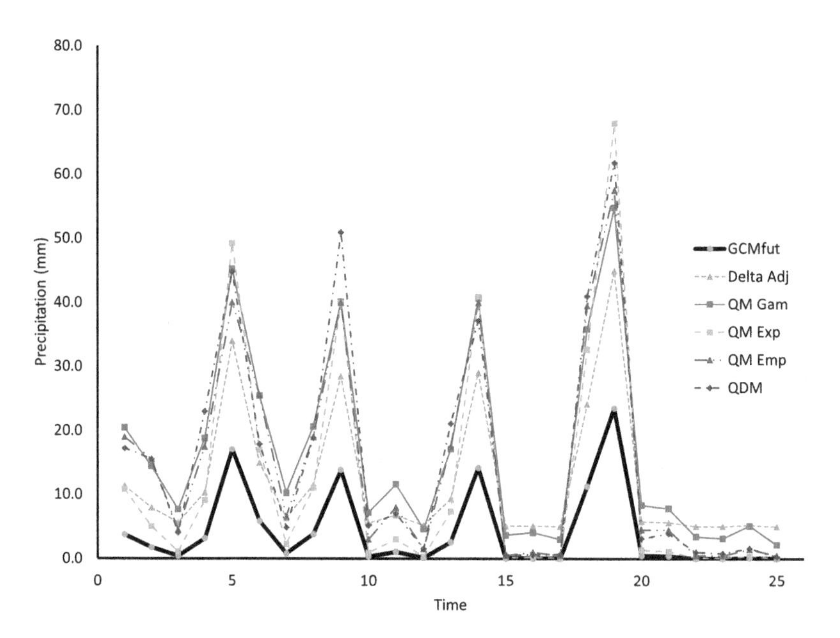

FIGURE 4.2 Time series of the original future GCM output data (black thick line with gray circle marker) and bias-corrected series: (1) Delta method (dotted light blue line with triangle marker), (2) QM with gamma distribution (solid blue line with square marker), (3) QM with exponential distribution (dashed green line with square marker), (4) QM with empirical CDF (dash-dotted blue line with triangle marker), (4) QDM with empirical CDF (dash-dotted red line with diamond marker).

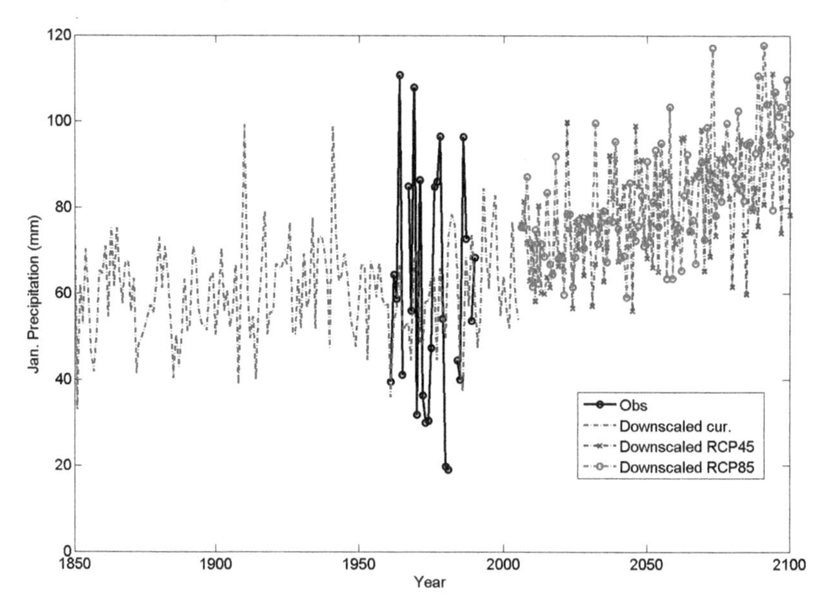

FIGURE 5.7 Downscaled January precipitation for the base period (dotted line) of 1850–2005 and the future period (dotted line with cross and circle markers for RCP4.5 and RCP8.5, respectively) of 2006–2100 employing the U wind speed and near-surface temperature of CGCM for the January average as well as the observed January precipitation (solid line with circles). See Tables 5.7 and 5.8 for detailed values.

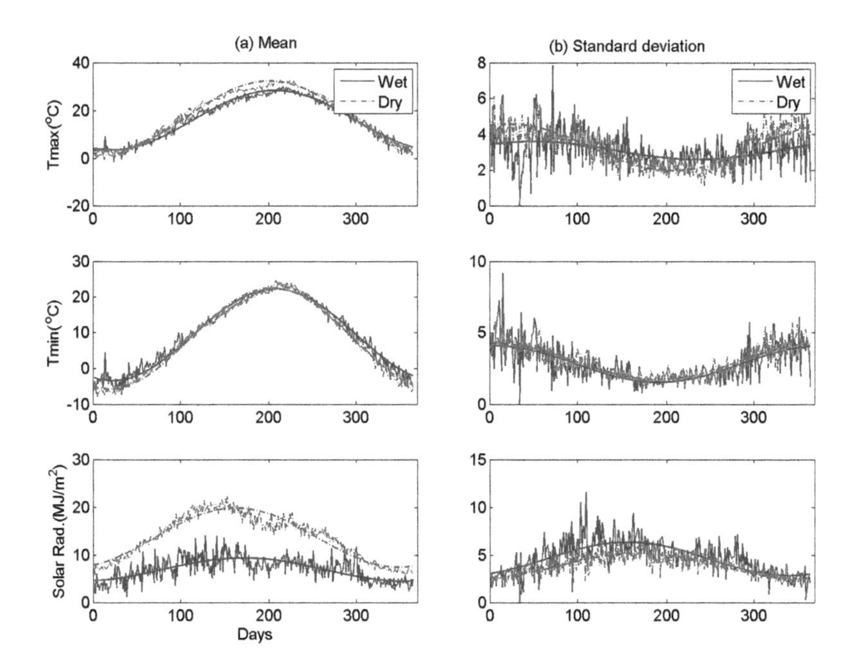

FIGURE 6.3 Mean (left column panels) and standard deviation (right column panels) of three weather variables (T_{max}, T_{min}, SR) for each day ($d = 1,\ldots,365$) regarding their occurrence condition (wet: solid line or dry: dash-dotted line) for Seoul as well as their fitted cosine function [thick smoothed lines overlapping with the original estimations from the data, see Eq. (6.21)]. Note that the parameters of the fitted cosine function in Eq. (6.21) were estimated with the Harmony Search algorithm presented in Chapter 2.

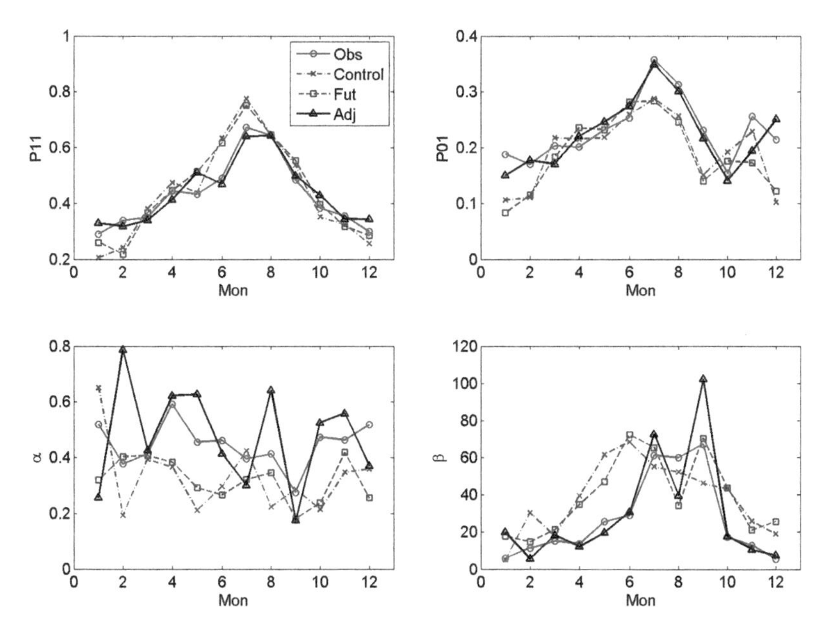

FIGURE 6.4 Adjusted parameters of the WGEN model for the RCP8.5 future scenario from the HadGem3-RA developed by Korea Meteorological Administration. Control run implies the HadGem3-RA dataset of 1979–2005, while the future scenario is 2006–2100.

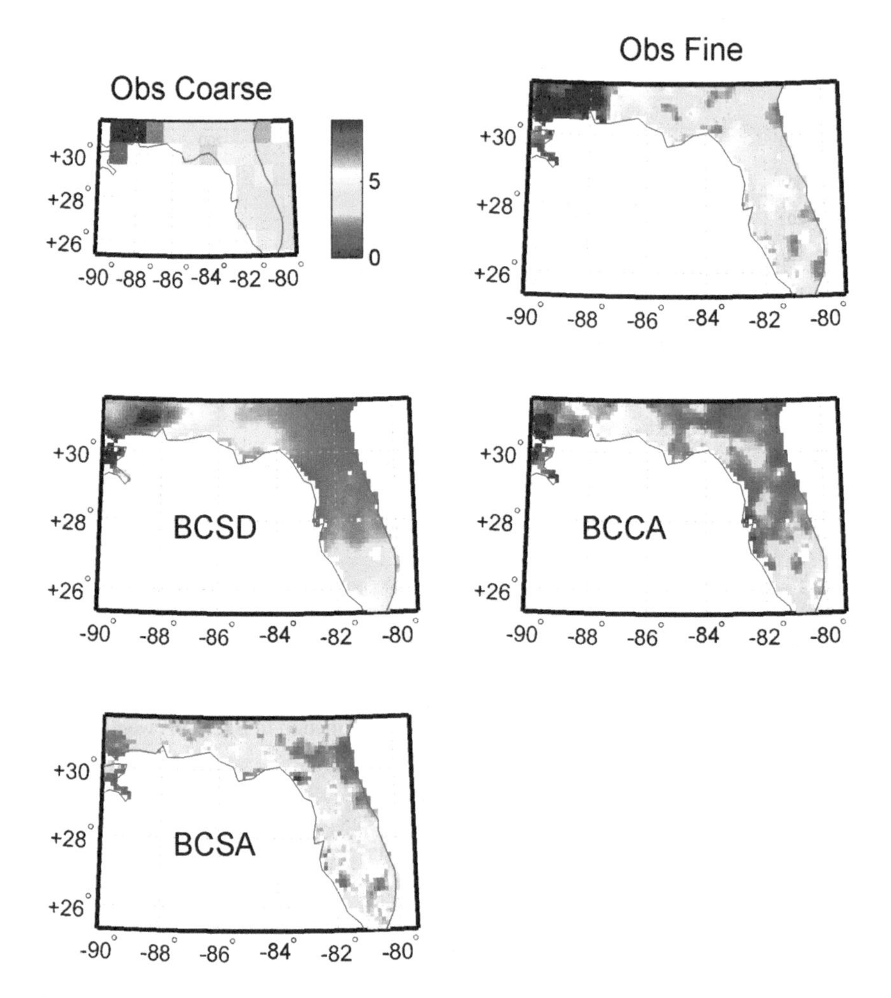

FIGURE 9.2 Example of spatial downscaling for monthly precipitation in a mean daily rate for coarse- and fine-scale data for Florida on May, 1990 with three BCSD, BCCA, and BCSA methods. The coarse and fine spatial observations are shown in the top left and right panels, respectively.

9 Spatial Downscaling

9.1 MATHEMATICAL BACKGROUND

9.1.1 Bilinear Interpolation

Assume that the z value for a_p and b_p [i.e., $z(a_p, b_p)$] needs to be found with the known values at the four points as $z(a_1, b_1)$, $z(a_1, b_2)$, $z(a_2, b_1)$, and $z(a_2, b_2)$ as shown in Figure 9.1. a and b are coordinates and $z(a, b)$ is the corresponding precipitation value for the coordinates as an example. At first, the z values of two points at (a_p, b_1) and (a_p, b_2) must be defined as

$$z(a_p,b_1) = \frac{a_2 - a_p}{a_2 - a_1} z(a_1,b_1) + \frac{a_p - a_1}{a_2 - a_1} z(a_2,b_1) \tag{9.1}$$

$$z(a_p,b_2) = \frac{a_2 - a_p}{a_2 - a_1} z(a_1,b_2) + \frac{a_p - a_1}{a_2 - a_1} z(a_2,b_2) \tag{9.2}$$

$$\begin{aligned} z(a_p,b_p) &\approx \frac{b_2 - b_p}{b_2 - b_1} z(a_p,b_1) + \frac{b_p - b_1}{b_2 - b_1} z(a_p,b_2) \\ &= \frac{1}{(a_2 - a_1)(b_2 - b_1)} \begin{bmatrix} a_2 - a_p & a_p - a_1 \end{bmatrix} \\ &\quad \begin{bmatrix} z(a_1,b_1) & z(a_1,b_2) \\ z(a_2,b_1) & z(a_2,b_2) \end{bmatrix} \begin{bmatrix} b_2 - b_p \\ b_p - b_1 \end{bmatrix} \end{aligned} \tag{9.3}$$

Example 9.1

Define the daily precipitation z value at $a_p = 4$, $b_p = 4$ [i.e., $z(4, 4)$] while $a_1 = 1$, $b_1 = 1$ and $a_2 = 5$, $b_2 = 5$. Assume the z values of $z(a_1, b_1) = 5$ mm, $z(a_1, b_2) = 10$ mm, $z(a_2, b_1) = 20$ mm, and $z(a_2, b_2) = 30$ mm are known. Then,

Solution:

$$z(4,4) = \frac{1}{(5-1)(5-1)} \begin{bmatrix} 5-4 & 4-1 \end{bmatrix} \begin{bmatrix} 5 & 10 \\ 20 & 30 \end{bmatrix} \begin{bmatrix} 5-4 \\ 4-1 \end{bmatrix} = 22.81\,\text{mm}$$

9.1.2 Nearest Neighbor Interpolation

The nearest neighbor interpolation simply selects the value of the nearest location. For example, the z values at $a_p = 4$, $b_p = 4$ for Example 9.1 [i.e., $z(4, 4)$] is assigned to be $z(a_p, b_p) = 30$ mm, since its location is the closest to $z(a_2, b_2)$.

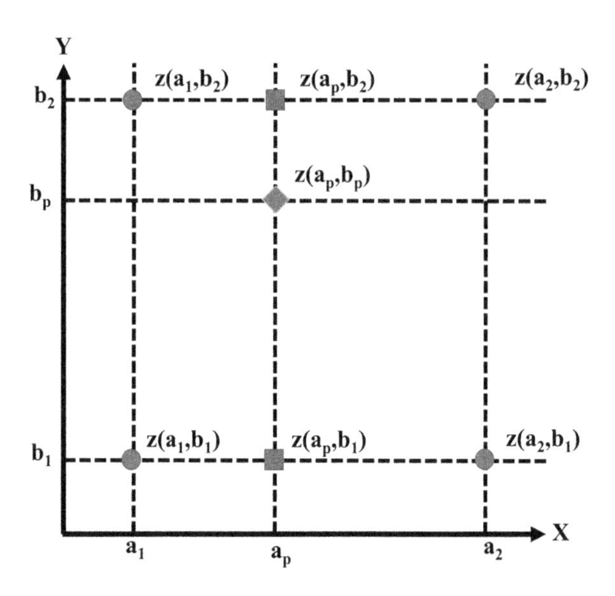

FIGURE 9.1 Example of bilinear interpolation. Here, a and b are coordinates and $z(a, b)$ is the corresponding precipitation value for the coordinate as an example.

9.2 BIAS CORRECTION AND SPATIAL DOWNSCALING

It has been reported that the bias correction and spatial downscaling (BCSD) model has been successfully used to downscale global climate model (GCM) results and to assess hydrological impacts of climate change (Salathé et al., 2007; Maurer and Hidalgo, 2008). The BCSD model utilizes observed climatic data (at both coarse and fine resolutions, e.g., $1° \times 1°$ and $1/8° \times 1/8°$, respectively) and GCM outputs with coarse resolution for the observation period and future period. The coarse GCM output (e.g., $1° \times 1°$) for the observation period is first bias-corrected using a quantile mapping (QM) technique based on the observed coarse data (Li et al., 2010; Thrasher et al., 2012; Grillakis et al., 2013; Maraun, 2013). Spatial downscaling is then performed to translate the bias-corrected GCM coarse output of the observation period to the GCM output of the prediction period via interpolation.

Let Y be the coarse grid data for a variable of interest, such as temperature or precipitation, while X denotes fine grid data. Note that (1) the uppercase letter denotes an unknown variable (e.g., X for an unknown fine grid variable) while the lowercase letter denotes a known variable (e.g., x for a known fine grid variable). Assume that GCM outputs of the original coarse resolution for future periods (denoted as $\mathbf{y}_{\mathrm{GCM}} = [y_{GCM,1}, \ldots, y_{GCM,q}]^T$) are known. The objective is to estimate the fine-resolution data of the GCM outputs $\mathbf{X}_{\mathrm{GCM}} = [X_{GCM,1}, \ldots, X_{GCM,p}]^T$ using the observed spatial pattern of the observed coarse and fine data, which are denoted as $\mathbf{y}_{\mathrm{obs}} = [y_{obs,1}, \ldots, y_{obs,q}]^T$ and $\mathbf{x}_{\mathrm{obs}} = [x_{obs,1}, \ldots, x_{obs,p}]^T$, respectively. Here, p and q refer to the number of grid cells for fine and coarse datasets, respectively.

1. Estimate the ratio of the variables of interest in the coarse resolution for each grid cell for all q numbers of GCM grid cells using

$$R_i = y_{GCM,i} / \mu^y_{obs,i} \quad \text{for} \quad i = 1,\ldots,q \tag{9.4}$$

$$\mu^y_{obs,i} = \frac{1}{H}\sum_{h=1}^{H} y^h_{obs,i} \tag{9.5}$$

where H is the number of observed years and $\mu^y_{obs,i}$ is the mean for the observed coarse data (y_{obs}) for the ith grid.
2. Interpolate the ratio (R_i) to the downscaling spatial resolution employing a spatial interpolation technique (e.g., nearest neighbor, bilinear, cubic, and spline) as in Example 9.1. The interpolated ratio is denoted as $\tilde{R}_j$, $j = 1, \ldots, p$. Note that the interpolation increases the number of cells from q (in the coarse resolution) to p (in the fine resolution).
3. Estimate the downscaled coarse data of the interested variable (i.e., $\mathbf{X}_{GCM}$) for all p numbers of downscaling grid cells by multiplying $\tilde{R}_j$ with the observed precipitation mean of the fine grid as follows:

$$X_{GCM,j} = \tilde{R}_j \times \mu^X_{obs,j} \quad \text{for } j = 1,\ldots,p \tag{9.6}$$

$$\mu^X_{obs,j} = 1/H \sum_{h=1}^{H} x^h_{obs,j} \tag{9.7}$$

Example 9.2

Downscale a 2×2 coarse grid dataset ($1° \times 1°$ resolution) into a 3×3 fine-resolution dataset ($1/2° \times 1/2°$ resolution) with the BCSD method with the following three datasets known as (1) observation coarse-resolution dataset (here 2×2); (2) observation fine-resolution dataset (here, 3×3); and (3) the target coarse dataset.

Solution:

In this example, gridded observed climate data of mean daily rates for January (mm/day) at 1/8° for fine resolution and at 1° for a coarse resolution over Florida for the current periods were drawn from Brekke et al. (2013). For simplicity, 1/2 degree resolution is considered instead of 1/8 degree, and the January dataset is employed. The coarse and fine-resolution datasets are presented in Tables 9.1 and 9.2, respectively, and their longitude and latitude are presented in the second and third rows of each table.

The coarse dataset consists of four grid cells for 29.5 and 30.5 latitudes as well as 277.5 and 278.5 longitudes. The fine-resolution dataset consists of nine grid cells for 28.563, 29.063, and 29.563 latitudes as well as 277.563, 278.063, and 278.563 longitudes. The observed period is 1971–1990 (i.e., 20 years), and it is assumed that the target period is 1991–1995. Note that the bias-correction part is omitted in this example by assuming that the coarse-resolution dataset is

TABLE 9.1
Example Data for the Coarse Resolution (1°) over Observation and Target Periods for Mean Daily Rates for January (mm/day) in Florida

		Y_1	Y_2	Y_3	Y_4
	Lat.	29.5	29.5	30.5	30.5
	Long.	277.5	278.5	277.5	278.5
Observation	1971	1.666	0.768	2.729	1.601
	1972	1.318	1.339	3.359	2.822
	1973	3.892	3.824	3.090	3.461
	1974	0.472	0.244	0.432	0.262
	1975	1.710	1.352	2.965	2.191
	1976	1.096	0.650	1.729	1.639
	1977	2.652	2.075	3.031	3.015
	1978	3.758	2.300	4.467	3.004
	1979	4.762	5.114	7.420	5.842
	1980	2.665	2.147	4.346	3.815
	1981	0.549	0.244	0.790	0.482
	1982	2.678	1.791	3.274	2.093
	1983	1.929	1.855	2.313	2.637
	1984	1.743	1.451	1.665	1.453
	1985	1.464	0.843	1.358	0.635
	1986	3.810	4.633	4.610	4.256
	1987	3.251	2.017	4.287	2.186
	1988	4.312	3.191	4.808	4.533
	1989	2.367	2.851	1.239	2.612
	1990	1.405	0.654	1.522	1.406
	Mean	2.375	1.967	2.972	2.497
	Std. dev.	1.252	1.378	1.695	1.423
Target	1991	2.711	2.214	5.796	3.267
	1992	1.606	1.266	3.824	2.345
	1993	3.398	3.994	3.534	4.067
	1994	5.729	4.444	6.928	6.022
	1995	2.659	1.788	2.863	1.633

bias-corrected with Section 4.4. The observed years are employed for verifying a downscaling method. The BCSD is as follows:

1. Estimate the ratios of the variables of interest [R_i in Eq. (9.4)] for the coarse resolution at each grid cell in Table 9.1. For example, for the year 1991 at Y_1

$$R_1 = 2.711 / 2.375 = 1.142$$

The total dataset of R_i is presented in Table 9.3

TABLE 9.2
Example Data for Fine Resolution (1/2 Degree) over Observation Period for Mean Daily Rates for January (mm/day) in Florida

Loc./Year	X_1	X_2	X_3	X_4	X_5	X_6	X_7	X_8	X_9
Lat	28.563	28.563	28.563	29.063	29.063	29.063	29.563	29.563	29.563
Long	277.563	278.063	278.563	277.563	278.063	278.563	277.563	278.063	278.563
1971	1.976	2.059	0.479	2.120	1.592	1.112	2.658	2.371	1.755
1972	1.216	1.505	1.151	2.168	2.276	1.847	4.075	3.355	3.299
1973	3.647	4.295	3.581	3.183	3.489	3.689	3.311	2.904	3.607
1974	0.755	0.241	0.166	0.305	0.140	0.203	0.243	0.348	0.232
1975	1.702	1.749	1.227	2.033	2.078	2.305	2.687	2.632	2.116
1976	1.503	1.208	0.336	1.494	1.642	1.268	1.416	1.748	2.016
1977	2.707	2.606	1.450	3.081	3.032	3.372	2.831	2.992	2.807
1978	4.165	2.737	1.910	3.811	3.278	2.646	4.998	4.104	2.862
1979	4.423	4.789	4.912	5.553	5.799	5.035	7.512	7.010	5.531
1980	2.592	2.105	1.899	4.095	3.582	2.216	4.939	4.594	4.327
1981	0.673	0.368	0.166	0.720	0.545	0.275	0.939	0.665	0.511
1982	2.772	2.705	1.512	2.312	2.125	2.306	4.139	2.111	1.897
1983	1.747	2.112	1.554	1.889	2.032	2.641	2.364	2.312	2.896
1984	1.703	1.549	1.558	1.795	1.558	1.482	1.196	1.433	1.552
1985	1.480	1.119	0.766	1.373	1.081	0.791	1.046	0.810	0.457
1986	3.511	4.930	5.320	4.290	4.851	5.579	4.769	4.368	3.443
1987	3.356	3.056	1.563	3.171	2.610	2.620	3.806	2.625	1.868
1988	4.909	3.444	3.060	5.111	4.432	3.805	4.343	4.784	4.912
1989	2.190	2.792	2.815	1.726	2.007	3.886	1.133	1.612	2.206
1990	1.965	1.017	0.375	1.703	1.482	1.100	1.528	1.621	1.576
Mean	2.450	2.319	1.790	2.597	2.481	2.409	2.997	2.720	2.493
Std. dev.	1.214	1.328	1.475	1.406	1.431	1.483	1.835	1.631	1.420

2. Interpolate the ratio (R_i) to the downscaling spatial resolution (here, 1/2° × 1/2°) employing a spatial interpolation technique (here, bilinear method).

   For example, for the year 1991 at the grid cell X_5

$$\tilde{R}_5 = \frac{1}{97.72 \times 111.20}\begin{bmatrix} 42.75 & 54.97 \end{bmatrix}\begin{bmatrix} 1.142 & 1.125 \\ 1.950 & 1.308 \end{bmatrix}\begin{bmatrix} 48.65 \\ 62.55 \end{bmatrix} = 1.389$$

   The total dataset of $\tilde{R}_j$, $j = 1, \ldots, p$, is presented in the second–sixth rows of Table 9.4 for the years 1991–1995. Note that the distance of two points in degree is estimated by the Haversine formula (Van Brummelen, 2012) as

$$d = 2r\arcsin\left\{\left[\sqrt{\sin^2\left(\frac{b_2 - b_1}{2}\right) + \cos(b_1)\cos(b_2)\sin^2\left(\frac{a_2 - a_1}{2}\right)}\right]\right\}$$

TABLE 9.3

Estimated R_i of Coarse Target Data in Table 9.1 for BCSD as in Eq. (9.4)

Year	Y_1	Y_2	Y_3	Y_4
1991	1.142	1.125	1.950	1.308
1992	0.676	0.643	1.287	0.939
1993	1.431	2.030	1.189	1.629
1994	2.412	2.259	2.331	2.411
1995	1.120	0.909	0.963	0.654

TABLE 9.4

Spatially Interpolated $\tilde{R}_j$ of Coarse Target Data in Table 9.2 with Bilinear Method and the Estimates of Y_{GCM} for BCSD

	Year	X_1	X_2	X_3	X_4	X_5	X_6	X_7	X_8	X_9
$\tilde{R}_j$	1991	1.189	1.574	2.101	1.161	1.389	1.759	1.282	1.405	1.686
	1992	0.711	1.007	1.386	0.685	0.902	1.201	0.748	0.924	1.191
	1993	1.452	1.327	1.384	1.747	1.581	1.637	2.219	1.998	2.058
	1994	2.399	2.365	2.633	2.329	2.354	2.670	2.561	2.639	3.036
	1995	1.096	1.015	1.072	0.988	0.882	0.901	1.018	0.878	0.867
$\mathbf{Y}_{GCM}$	1991	2.912	3.650	3.760	3.015	3.448	4.238	3.843	3.821	4.203
	1992	1.742	2.334	2.482	1.778	2.237	2.892	2.241	2.514	2.970
	1993	3.558	3.077	2.478	4.537	3.924	3.943	6.649	5.434	5.131
	1994	5.875	5.486	4.712	6.048	5.842	6.433	7.673	7.178	7.571
	1995	2.686	2.354	1.919	2.565	2.189	2.171	3.051	2.388	2.162

where r is the radius of the sphere (r = 6,371 km) and b_i and a_i are the latitude and longitude of two points i, i = 1,2 in radians (converted by multiplying by $\pi/180$). For example, $[b_1$ and $a_1]$ = [29.5 277.5] × $\pi/180$ = [0.5149 4.8433] and $[b_2$ and $a_2]$ = [29.5 278.5] × $\pi/180$ = [0.5149 4.8607]

$$d = 2\times 6371\times \arcsin\left\{\left[\sqrt{\sin^2(0)+\cos(0.5149)\cos(0.5149)\sin^2\left(\frac{4.8607-4.8433}{2}\right)}\right]\right\}$$

$$= 96.779\,\text{km}$$

3. Multiply the interpolated ratios $\tilde{R}_j$ with the observed precipitation mean of the fine grid to estimate $X_{GCM,j}$ j = 1, … , p. For example, the grid cell Y_5 for the year 1991,

$$X_{GCM,5} = \tilde{R}_5 \times \mu^X_{obs,5} = 1.389\times 2.481 = 3.448$$

The total dataset of $\tilde{R}_j$, j = 1, … , p (here, p = 9) is presented in the seventh–eleventh rows of Table 9.4 for the years 1991–1995.

9.3 BIAS CORRECTION AND CONSTRUCTED ANALOGUES

The bias correction and constructed analogue (BCCA) method obtains the spatial information from a linear combination of fine-scale historical analogues for downscaling the coarse-scale GCM outputs. The BCCA approach was described by Hidalgo et al. (2008) and Maurer et al. (2010). The BCCA method involves the following conditions and procedures: (i) A new pattern for coarse resolution is known, but the fine-resolution pattern is unknown; (ii) a subset of patterns from a historical library is used based on the presence of spatial similarities; and (iii) a linear combination of predictor patterns (of fine resolution) produces a least-square fit analogue.

Assume that the known coarse-resolution data $y_{GCM} = \left[y_{GCM,1},\ldots,y_{GCM,q}\right]^T$ needs to be downscaled. The following procedure is used:

1. Find the K closest patterns from the observed pattern library with a target future GCM of RMSE over the q number of coarse grid cells as

$$RMSE_h = \sqrt{\frac{\sum_{i=1}^{q}\left(y_{GCM,i} - y_{obs,i}^{h}\right)^2}{q}} \tag{9.8}$$

where h represents the observed year for $h = 1, \ldots, H$, and H is the number of observed years (i.e., base period).

2. Select the K years whose $RMSE_h$ are the K smallest among the H observed years. Set $(k) = (1), \ldots, (K)$ for the time indices of the K smallest $RMSE_h$ and collect the corresponding coarse observations as $y_{obs}^{(k)}$. Usually, $K = 30$ has been suggested (Hidalgo et al., 2008).
3. Estimate the multiple linear regression coefficient vector $\mathbf{A}$ of the least-square fit (see Eq. (5.12) for further detail) as follows:

$$\mathbf{A} = \left[\left(\vec{\mathbf{y}}_{obs}^{T}\vec{\mathbf{y}}_{obs}\right)^{-1}\vec{\mathbf{y}}_{obs}^{T}\right]\mathbf{y}_{GCM}^{T} \tag{9.9}$$

where $\mathbf{y}_{\mathrm{GCM}}$ is the $q \times 1$ vector for the target coarse-resolution data and $\vec{\mathbf{y}}_{obs}$ is the $q \times (K + 1)$ matrix whose elements are the selected GCM outputs of the observed period from step (2); i.e., $y_{obs,i}^{(k)}$ $k = 1, \ldots, K$, where the kth-order coarse observations includes vector $\mathbf{1}$ of the first column, i.e.,

$$\vec{\mathbf{y}}_{obs} = \begin{bmatrix} 1 & y_{obs,1}^{(1)} & \cdots & y_{obs,1}^{(K)} \\ 1 & y_{obs,2}^{(1)} & \cdots & y_{obs,2}^{(K)} \\ \vdots & \vdots & \cdots & \vdots \\ 1 & y_{obs,q}^{(1)} & \cdots & y_{obs,q}^{(K)} \end{bmatrix} \text{ and } \mathbf{y}_{GCM} = \begin{bmatrix} y_{GCM,1} \\ y_{GCM,2} \\ \vdots \\ y_{GCM,q} \end{bmatrix}$$

Note that $\mathbf{A}$ is the $(K + 1) \times 1$ parameter vector.

4. Estimate the downscaled fine-resolution data by applying the linear regression coefficient vector **A** as

$$\mathbf{X_{GCM}} = \vec{\mathbf{x}}_{\mathbf{obs}}\mathbf{A} \tag{9.10}$$

where $\vec{\mathbf{x}}_{\mathbf{obs}}$ is the $p \times (K+1)$ matrix, whose elements are the selected fine observed data as

$$\vec{\mathbf{x}}_{\mathbf{obs}} = \begin{bmatrix} 1 & x_{obs,1}^{(1)} & \cdots & x_{obs,1}^{(K)} \\ 1 & x_{obs,2}^{(1)} & \cdots & x_{obs,2}^{(K)} \\ \vdots & \vdots & \cdots & \vdots \\ 1 & x_{obs,p}^{(1)} & \cdots & x_{obs,p}^{(K)} \end{bmatrix} \tag{9.11}$$

Example 9.3

Spatially downscale the same dataset as in Example 9.2 but with BCCA instead of BCSD. It is assumed that $K = 2$ and the coarse-resolution data for the year 1991 (i.e., $\mathbf{y}_{GCM}$ = [2.711, 2.214, 5.796, 3.267]) are downscaled into the nine high-resolution grid cells:

Solution:

1. Find the K closest patterns of the observed pattern library with a target future GCM of RMSE over the q number of coarse grid cells. For example, the RMSE for the year 1971 between the known coarse GCM and the coarse observation in Table 9.1 is

$$RMSE_1 = \sqrt{\frac{\sum_{i=1}^{4}\left(y_{GCM,i} - y_{obs,i}^{(1)}\right)^2}{4}}$$

$$= \sqrt{\frac{(2.711-1.666)^2 + \cdots + (3.267-1.601)^2}{4}} = 1.960$$

All the estimated RMSEs for the years 1971–1990 are shown in the third column of Table 9.5. The two smallest RMSEs are for the years 1980 and 1978, as indicated in the tenth and eighth rows of the second column in Table 9.5.

2. Fit the multiple linear regression model to the selected coarse-resolution data for the years 1980 and 1978 from Table 9.1 and estimate the parameter vector **A** as

$$\vec{\mathbf{y}}_{obs} = \begin{bmatrix} 1 & 2.665 & 3.758 \\ 1 & 2.147 & 2.300 \\ 1 & 4.346 & 4.467 \\ 1 & 3.815 & 3.004 \end{bmatrix} \text{ and } \mathbf{y}_{GCM} = \begin{bmatrix} 2.711 \\ 2.214 \\ 5.796 \\ 3.267 \end{bmatrix}$$

TABLE 9.5
Estimated RMSE for the BCCA Downscaling of the January Mean Daily Precipitation Rate (mm/day) in Florida for the Year 1991 in the Bottom Part of Table 9.1

Ind	Years	RMSE	Sort_Ind	Sort RMSE
1	1971	1.960	10	0.776
2	1972	1.487	8	0.857
3	1973	1.684	17	0.971
4	1974	3.416	18	1.235
5	1975	1.652	7	1.390
6	1976	2.462	12	1.407
7	1977	1.390	2	1.487
8	1978	0.857	16	1.537
9	1979	2.339	5	1.652
10	1980	0.776	3	1.684
11	1981	3.216	13	1.821
12	1982	1.407	1	1.960
13	1983	1.821	19	2.330
14	1984	2.339	14	2.338
15	1985	2.741	9	2.339
16	1986	1.537	6	2.462
17	1987	0.971	20	2.543
18	1988	1.235	15	2.741
19	1989	2.330	11	3.216
20	1990	2.543	4	3.416

Note that the third and fourth columns are the sorted RMSE and its corresponding index, respectively. The tenth and eighth years (1980 and 1978) present two lowest RMSE.

$$\mathbf{A} = \left[\left(\vec{\mathbf{y}}_{obs}{}^{T}\vec{\mathbf{y}}_{obs}\right)^{-1}\vec{\mathbf{y}}_{obs}{}^{T}\right]\mathbf{y}_{GCM}{}^{T} = \begin{bmatrix} -1.921 \\ 0.943 \\ 0.697 \end{bmatrix}$$

3. Estimate the downscaled data by applying the coefficient vector **A** and the fine resolution data corresponding to the selected index in step (1) (i.e., the fine-resolution data in Table 9.6 for the years 1980 and 1978) as in Eq. (9.10):

$$\mathbf{X}_{\text{GCM}} = \vec{\mathbf{x}}_{\text{obs}}\mathbf{A} = \begin{bmatrix} 1 & 2.592 & 4.165 \\ 1 & 2.105 & 2.737 \\ \vdots & \vdots & \vdots \\ 1 & 4.327 & 2.862 \end{bmatrix}\begin{bmatrix} -1.921 \\ 0.943 \\ 0.697 \end{bmatrix} = \begin{bmatrix} 3.429 \\ 1.973 \\ \vdots \\ 4.156 \end{bmatrix}$$

4. Repeat steps (1)–(3) for all the years until all the fine-resolution data are estimated. The downscaled dataset with BCCA is presented in Table 9.7.

TABLE 9.6
Estimated Parameters by a Multiple Linear Regression Fit in Eq. (9.9) with Two Lowest RMSE Year (1980 and 1978, see Table 9.5) as well as Their Relative Fine-Resolution Dataset for the BCCA Downscaling of the January Mean Daily Precipitation Rate (mm/day) for the Year 1991 at the Bottom Part of Table 9.1

	Param	X_1	X_2	X_3	X_4	X_5	X_6	X_7	X_8	X_9
b_0	−1.921	1	1	1	1	1	1	1	1	1
b_1 (1980)	0.943	2.592	2.105	1.899	4.095	3.582	2.216	4.939	4.594	4.327
b_2 (1978)	0.697	4.165	2.737	1.910	3.811	3.278	2.646	4.998	4.104	2.862

TABLE 9.7
Downscaled Fine-Resolution Dataset of the January Mean Daily Precipitation Rate (mm/day) in Florida for the Years of 1991–1995 (See the Bottom Part of Table 9.1 for the Coarse-Resolution Dataset)

Year	X_1	X_2	X_3	X_4	X_5	X_6	X_7	X_8	X_9
1991	3.429	1.973	1.202	4.599	3.744	2.015	6.223	5.274	4.156
1992	1.708	1.775	0.920	2.226	2.297	2.689	3.251	3.182	2.326
1993	3.326	4.322	4.299	3.600	4.010	4.477	3.905	3.549	3.275
1994	6.242	4.655	4.246	6.754	6.037	5.129	6.361	6.742	6.520
1995	2.829	2.718	1.428	2.053	1.785	1.904	4.104	1.477	1.467

9.4 BIAS CORRECTION AND STOCHASTIC ANALOGUE

The bias correction and stochastic analogue (BCSA) technique generates replicates for downscaling the coarse data to fine scale data by employing the correlation coefficients of the standard normal variable that is transformed from the QM of observed fine data (Hwang and Graham, 2013).

The BCSA technique involves the following conditions and procedures: (i) QM is conducted to convert observed fine-resolution precipitation into a standard normal variable for each grid (bias correction); (ii) spatially correlated fine precipitation replicates are generated and back-transformed to their observed empirical distributions employing the correlation coefficient of the transformed standard normal variable; and (iii) the replicate that the spatial mean is closest to the spatial mean of target coarse data is selected from the replicates of generated fine precipitation data. The detailed description is as follows:

1. Transform the coarse and fine-resolution dataset into standard normal variables as a bias-correction process with QM like in Eq. (4.7). However, there is a slight difference from the original QM in Eq. (4.7). The standard normal variables can be obtained by

$$\tilde{x}_{obs,i} = \Phi^{-1}(F_{xo}(x_{obs,i})) \tag{9.12}$$

where F_{xo} is the fitted Cumulative Distribution Function (CDF) of an assumed probability distribution (e.g., gamma, exponential, and normal) to the fine-resolution observed data of the ith grid cell ($x_{obs,i}$; $i = 1, \ldots ,p$) and Φ^{-1} is the inverse function of the standard normal CDF in Eq. (2.19). Also, the future GCM projection of the fine-resolution data must be transformed into standard normal variable as

$$\tilde{y}_{GCM,j} = \Phi^{-1}\left(F_{yo}\left(y_{GCM,j}\right)\right) \quad (9.13)$$

where F_{yo} is the CDF fitted to the coarse-resolution observed data of the jth grid cell ($y_{obs,j}$; $j = 1, \ldots ,q$). Note that F_{yo} should be fitted to the observed data (i.e., not $y_{GCM,j}$ but $y_{obs,j}$).

2. Estimate the correlation coefficient matrix between the transformed observed data of the fine resolution as

$$\rho_{ij} = corr\left(\tilde{x}_{obs,i}, \tilde{x}_{obs,j}\right) \quad \text{for } i, j = 1, \ldots, p \quad (9.14)$$

where p is the number of fine-resolution grid cells.

3. Simulate the N_G number of replicates from the multivariate normal distribution as

$$\tilde{\mathbf{X}}_{\mathbf{n}} \sim \mathbf{MVN}(\mathbf{0}, \boldsymbol{\rho}) \quad (9.15)$$

where $\mathbf{0}$ is the $p \times 1$ zero vector, and where $\boldsymbol{\rho}$ is the $p \times p$ correlation matrix estimated from step 1. $\tilde{\mathbf{X}}_{\mathbf{n}}$ is the candidate that mimics the spatial correlations of observed fine-resolution data.

4. Select one among N replicates, whose spatial mean is the closest to the future spatial mean as follows:

$$\tilde{\mathbf{X}}_{\mathbf{GCM}} = \min_{\tilde{\mathbf{X}}_{\mathbf{n}}, n \in \mathbf{N}} \left(\left|\mu_{\mathbf{n}}^{\tilde{\mathbf{X}}} - \mu_{\mathbf{Y}_{\mathbf{GCM}}}\right|\right) \quad (9.16)$$

where $\mu_{\mathbf{n}}^{\tilde{\mathbf{X}}}$ is the spatial mean vector of $\tilde{\mathbf{X}}_{\mathbf{n}}$ and $\mu_{\mathbf{Y}_{\mathbf{GCM}}}$ is the spatial mean vector of target future GCM.

5. Back-transform the selected standard normal variable of fine-resolution data $\tilde{\mathbf{X}}_{\mathbf{GCM}}$ into the original domain for each cell as

$$X_{GCM,i} = F_{xo}^{-1}(\Phi(\tilde{x}_{GCM,i})), \quad i = 1, \ldots, p$$

where p is the number of fine grid cells.

6. Repeat steps (4)–(5) until all the target coarse-resolution data are downscaled.

Example 9.4

Spatially downscale the same dataset as in Example 9.2 but with BCSA instead of BCSD. The same dataset of the January mean daily precipitation rate (unit: mm/day) in Florida as Example 9.2 is employed.

Solution:

The target coarse-resolution data for the year 1991 (i.e., $\mathbf{y}_{GCM}$ = [2.711, 2.214, 5.796, 3.267]) is downscaled into nine high-resolution grid cells:

1. Fit the coarse and fine observed data into an assumed distribution. Here, the gamma distribution [see Eqs. (2.25) and (2.26) for its Probability Distribution Function (PDF) and CDF] is used, since it has been popularly employed to fit precipitation data. The parameters are estimated with method of moments (MOMs) as in Eqs. (2.29) and (2.30).

   For example, the mean and standard deviation of $y_{obs,1}$ are 2.375 and 1.252, respectively, as shown in Table 9.1. The parameters of the gamma distribution with MOM are $\beta = \frac{\sigma^2}{\mu} = \frac{1.252^2}{2.375} = 0.660$ and $\alpha = \frac{\mu}{\beta} = \frac{2.375}{0.660} = 3.600$ [see Eqs. (2.29) and (2.30), respectively]. All the estimated parameters of observed coarse and fine data are shown in the upper part of Tables 9.8 and 9.9, respectively. The future coarse-resolution data of 1991 is transformed into the standard normal variable as

   $$\tilde{y}_{GCM,1} = \Phi^{-1}\left\{F_{yo}(y_{GCM,1};\, \alpha,\beta)\right\} = \Phi^{-1}\left\{F_{yo}(2.711;\, 3.600, 0.660)\right\} = 0.432$$

   All the coarse and fine data transformed to the standard normal variable are presented at the lower part of Tables 9.8 and 9.9, respectively.
2. Estimate the correlation coefficient matrix in Eq. (9.14) with observed fine-resolution data (i.e., $x_{obs,i}$, $i = 1, \ldots, p$) by Eq. (2.59). The estimated correlation matrix is shown in Table 9.10.

TABLE 9.8

Parameters of the Gamma Distribution Fitted to the Coarse Observed Data for Each Cell as well as Their Transformed Coarse Data ($y_{GCM,i}$; $i = 1, \ldots, q$)

	Parameter	Y_1	Y_2	Y_3	Y_4	
Parameters of gamma distribution	α	3.6	2.038	3.072	3.082	
	β	0.66	0.965	0.967	0.81	
	Year	Y_1	Y_2	Y_3	Y_4	**Mean**
Transformed coarse data to standard normal	1991	0.432	0.405	1.497	0.680	0.754
	1992	−0.516	−0.344	0.649	0.083	−0.032
	1993	0.894	1.367	0.501	1.115	0.969
	1994	2.119	1.563	1.903	1.985	1.892
	1995	0.393	0.103	0.127	−0.499	0.031

TABLE 9.9
Parameters of the Gamma Distribution Fitted to the Coarse Observed Data for Each Cell as well as Their Transformed Coarse Data ($y_{GCM,i}$; $i = 1, \ldots, q$) with MOM

	Parameters	X_1	X_2	X_3	X_4	X_5	X_6	X_7	X_8	X_9
Param. of Gam.	α_x	4.075	3.051	1.472	3.412	3.007	2.64	2.667	2.782	3.085
	β_x	0.601	0.76	1.216	0.761	0.825	0.912	1.124	0.978	0.808
	QM X	X_1	X_2	X_3	X_4	X_5	X_6	X_7	X_8	X_9
Transf. coarse data to std. normal	1971	−0.250	−0.010	−1.017	−0.178	−0.517	−0.896	0.015	−0.020	−0.387
	1972	−1.093	−0.507	−0.213	−0.139	0.047	−0.202	0.729	0.561	0.702
	1973	1.021	1.381	1.212	0.568	0.814	0.944	0.370	0.311	0.876
	1974	−1.811	−2.679	−1.779	−2.730	−3.196	−2.640	−2.689	−2.346	−2.800
	1975	−0.522	−0.275	−0.146	−0.249	−0.103	0.136	0.033	0.147	−0.087
	1976	−0.740	−0.827	−1.292	−0.746	−0.471	−0.726	−0.873	−0.479	−0.166
	1977	0.370	0.398	0.035	0.504	0.550	0.780	0.114	0.361	0.402
	1978	1.334	0.487	0.356	0.933	0.695	0.360	1.110	0.931	0.436
	1979	1.480	1.619	1.723	1.776	1.892	1.558	1.959	2.056	1.789
	1980	0.280	0.027	0.349	1.085	0.866	0.074	1.087	1.150	1.247
	1981	−1.970	−2.260	−1.777	−1.760	−1.893	−2.377	−1.370	−1.689	−2.001
	1982	0.419	0.466	0.082	−0.028	−0.066	0.137	0.757	−0.201	−0.264
	1983	−0.475	0.032	0.113	−0.372	−0.139	0.357	−0.163	−0.060	0.459
	1984	−0.521	−0.463	0.116	−0.456	−0.549	−0.517	−1.084	−0.756	−0.575
	1985	−0.766	−0.934	−0.611	−0.875	−1.063	−1.307	−1.244	−1.467	−2.123
	1986	0.934	1.685	1.862	1.186	1.490	1.776	1.020	1.051	0.785
	1987	0.832	0.693	0.119	0.560	0.281	0.343	0.608	0.143	−0.289
	1988	1.742	0.925	0.980	1.580	1.296	1.002	0.845	1.231	1.522
	1989	−0.055	0.524	0.862	−0.519	−0.158	1.042	−1.150	−0.594	−0.018
	1990	−0.260	−1.064	−1.209	−0.541	−0.623	−0.909	−0.774	−0.586	−0.551

TABLE 9.10
Correlation Coefficient of the Fine-Resolution Observed Data

	X_1	X_2	X_3	X_4	X_5	X_6	X_7	X_8	X_9
X_1	1.000	0.901	0.816	0.917	0.879	0.840	0.793	0.818	0.771
X_2	0.901	1.000	0.922	0.899	0.929	0.957	0.815	0.836	0.832
X_3	0.816	0.922	1.000	0.831	0.872	0.936	0.722	0.777	0.786
X_4	0.917	0.899	0.831	1.000	0.978	0.866	0.926	0.956	0.902
X_5	0.879	0.929	0.872	0.978	1.000	0.927	0.908	0.953	0.936
X_6	0.840	0.957	0.936	0.866	0.927	1.000	0.768	0.837	0.869
X_7	0.793	0.815	0.722	0.926	0.908	0.768	1.000	0.951	0.863
X_8	0.818	0.836	0.777	0.956	0.953	0.837	0.951	1.000	0.951
X_9	0.771	0.832	0.786	0.902	0.936	0.869	0.863	0.951	1.000

3. Simulate the N_G number of replicates from the multivariate normal distribution with zero mean and correlation matrix estimated from step (2) in Table 9.10. Here, $N_G = 30$ is employed for simplicity, but $N \geq 10^4$ is recommended in real practice. The simulated data is shown in Table 9.11.
4. Calculate the distance between the spatial mean of the N_G replicates simulated from step (3) (presented in the last column of Table 9.11) and the same of the future coarse data (presented in the last column of the below part of Table 9.8) for each year. For example, the spatial mean of the future coarse data in 1991 is 0.754 and its distance to the one of 30 replicates is shown in Table 9.12 (e.g., dist = |0.754–0.860| = 0.1059).

TABLE 9.11
Simulated Data from Multivariate Normal Distribution with Correlations in Table 9.10

	X_1	X_2	X_3	X_4	X_5	X_6	X_7	X_8	X_9	Mean
1	0.803	0.626	0.571	1.089	1.091	0.490	1.251	0.899	0.917	0.860
2	−1.320	−1.365	−1.075	−0.985	−1.051	−0.991	−1.082	−0.933	−0.514	−1.035
3	−0.274	−0.017	0.498	−0.946	−0.690	0.004	−1.427	−1.390	−1.132	−0.597
4	0.272	−0.193	0.026	−0.108	0.002	0.123	−0.257	−0.299	−0.258	−0.077
5	1.490	1.815	1.575	1.867	2.060	1.843	2.464	1.819	1.772	1.856
6	1.437	2.070	2.173	1.776	2.097	2.293	1.552	1.904	2.388	1.966
7	−0.028	−0.157	−0.248	−0.081	−0.240	−0.689	−0.289	−0.045	0.084	−0.188
8	0.924	0.828	0.065	0.957	0.794	0.744	0.914	1.334	1.318	0.875
9	−0.321	−0.069	−0.947	0.138	0.167	−0.159	0.329	0.199	0.053	−0.068
10	0.661	1.118	0.399	0.820	0.915	1.243	0.788	0.894	1.140	0.887
11	1.915	1.952	1.697	1.563	1.370	1.775	1.469	1.288	1.040	1.563
12	0.157	0.314	0.061	0.129	−0.030	−0.026	0.189	0.015	0.145	0.106
13	−0.301	−0.480	−0.731	−0.451	−0.382	−0.579	−0.585	−0.715	−0.922	−0.572
14	−0.500	−0.551	−0.492	−0.172	−0.319	−0.931	−0.092	−0.614	−0.820	−0.499
15	0.716	0.912	0.596	0.892	1.207	1.167	0.424	0.835	1.517	0.918
16	1.337	1.936	1.575	1.691	1.557	1.507	1.523	1.230	0.879	1.471
17	2.126	2.162	2.214	2.114	2.186	2.402	1.310	1.484	1.486	1.943
18	0.054	−0.475	−0.176	−0.264	−0.415	−0.598	−0.577	−0.075	−0.285	−0.312
19	0.163	0.335	1.257	0.308	0.317	0.680	−0.276	0.117	0.160	0.340
20	−0.633	−0.610	−0.440	−0.509	−0.445	−0.444	−0.700	−0.430	−0.329	−0.504
21	1.612	1.346	1.232	0.932	0.778	0.916	0.560	0.694	0.640	0.968
22	−0.075	−0.163	−0.377	−0.056	0.011	0.053	−0.458	−0.431	−0.453	−0.217
23	−0.473	−0.808	−0.841	−0.252	−0.391	−0.732	−0.174	−0.072	−0.150	−0.433
24	2.184	1.828	1.624	1.907	1.559	1.356	1.796	1.682	1.155	1.677
25	0.810	0.389	−0.353	1.133	0.567	0.087	0.960	0.984	0.437	0.557
26	0.716	0.487	0.328	0.791	0.647	0.486	1.102	0.966	0.548	0.674
27	−1.006	−0.855	−0.481	−1.034	−0.794	−0.536	−1.571	−1.234	−0.898	−0.934
28	0.434	0.467	0.210	0.357	0.383	0.445	0.205	0.411	1.300	0.468
29	0.520	0.375	0.521	0.354	0.330	0.416	−0.292	−0.106	0.784	0.323
30	−1.092	−1.050	−0.575	−0.914	−0.789	−0.700	−1.008	−0.933	−0.504	−0.841

TABLE 9.12
Distance between the Spatial Mean of Standard Normal Future Coarse 1991 Data in Table 9.8 and the One of Simulated Data in Table 9.11

Order	Distance	Order	Distance	Order	Distance
1	0.1059	11	0.8095	21	0.2143
2	1.7888	12	0.6476	22	0.9703
3	1.3508	13	1.3254	23	1.1863
4	0.8306	14	1.2526	24	0.9232
5	1.1025	15	0.1647	25	0.1967
6	1.2119	16	0.7169	26	**0.0792**
7	0.9417	17	1.1889	27	1.6879
8	0.1217	18	1.0659	28	0.2858
9	0.8213	19	0.4135	29	0.4311
10	0.1329	20	1.2581	30	1.5942

Note that the absolute difference is used, since the negative and positive differences are the same.

5. Find the smallest distance for the future coarse data for 1991, which is with the 26th replicate in Table 9.11. This replicate is selected for the future fine-resolution data for the year 1991. All the selected replicate (i.e., $\tilde{\mathbf{X}}_{\mathbf{GCM}}$) for the years 1991–1995 are displayed in Table 9.13 and the number of selected replicates is shown in the last column of this table.
6. Back-transform the selected standard normal variable of the fine-resolution data in Table 9.13 into the original domain for each cell. For example, the year 1991 and the first fine-resolution cell (i.e., $X_{GCM,1}$), whose value is 0.716 as

$$X_{GCM,1} = F_{xo}^{-1}(\Phi(0.716); 4.075, 0.601) = 3.185$$

TABLE 9.13
Selected Replicates with the Lowest Distances of the Spatial Means between the Replicates in Table 9.11 and the Standard Normal Future Coarse Data in Table 9.8

	X_1	X_2	X_3	X_4	X_5	X_6	X_7	X_8	X_9	Rep. #
1991	0.716	0.487	0.328	0.791	0.647	0.486	1.102	0.966	0.548	26
1992	−0.321	−0.069	−0.947	0.138	0.167	−0.159	0.329	0.199	0.053	9
1993	1.612	1.346	1.232	0.932	0.778	0.916	0.560	0.694	0.640	21
1994	1.490	1.815	1.575	1.867	2.060	1.843	2.464	1.819	1.772	5
1995	0.157	0.314	0.061	0.129	−0.030	−0.026	0.189	0.015	0.145	12

TABLE 9.14

Downscaled Fine-Resolution Data of the January Mean Daily Precipitation Rate in Florida from the Coarse Data Shown in Table 9.1 with BCSA through the Back-Transfromation of Table 9.13 to the Original Domain Using the Gamma Distribution and the Parameter Set of Table 9.9

	X_1	X_2	X_3	X_4	X_5	X_6	X_7	X_8	X_9
1991	3.185	2.737	1.866	3.559	3.195	2.852	4.978	4.180	3.040
1992	1.902	1.987	0.521	2.536	2.443	1.903	3.231	2.717	2.301
1993	4.664	4.226	3.629	3.811	3.423	3.634	3.703	3.615	3.193
1994	4.440	5.219	4.500	5.768	6.229	5.753	9.320	6.316	5.490
1995	2.438	2.486	1.484	2.523	2.173	2.077	2.966	2.425	2.429

Here, the gamma distribution is employed with MOM parameter estimation as shown in step (1), and their parameter set for each set from fine observations in Table 9.9. All the back-transformed data $X_{GCM,i}$ $i = 1, \ldots, p$ are shown in Table 9.14.

9.5 SUMMARY AND COMPARISON

Three spatial downscaling models are described in this study with the example for the Florida monthly precipitation with mean daily rate (mm/day) as BCSD, BCCA, and BCSA. Since each model employs a different statistical technique, its output also has unique statistical characteristics and a caution must be exercised to employ in application.

In Figure 9.2, one example of spatial downscaling for monthly precipitation in average daily rate (mm/day) is presented for the entire Florida region for May 1990. Its original coarse and fine observations are presented at the left and right of the top panels, respectively. The BCSD results present its smoothed values through the region, because the method employs a spatial smoothing interpolation technique, while the other two present rough surfaces. Note that the BCSD model has a better prediction capability than the others, such that its value is close to true value. However, due to high smoothing characteristics, the model might not produce extreme characteristics and caution needs to be exercised for extreme hydrological studies, such as floods. Meantime, the other two models well preserve the extreme characteristics. However, BCSA also lacks predictive capability, and hence the simulation is rather random and spatial distribution of precipitation is not well matched with coarse (left top panel) and fine (left bottom panel) data.

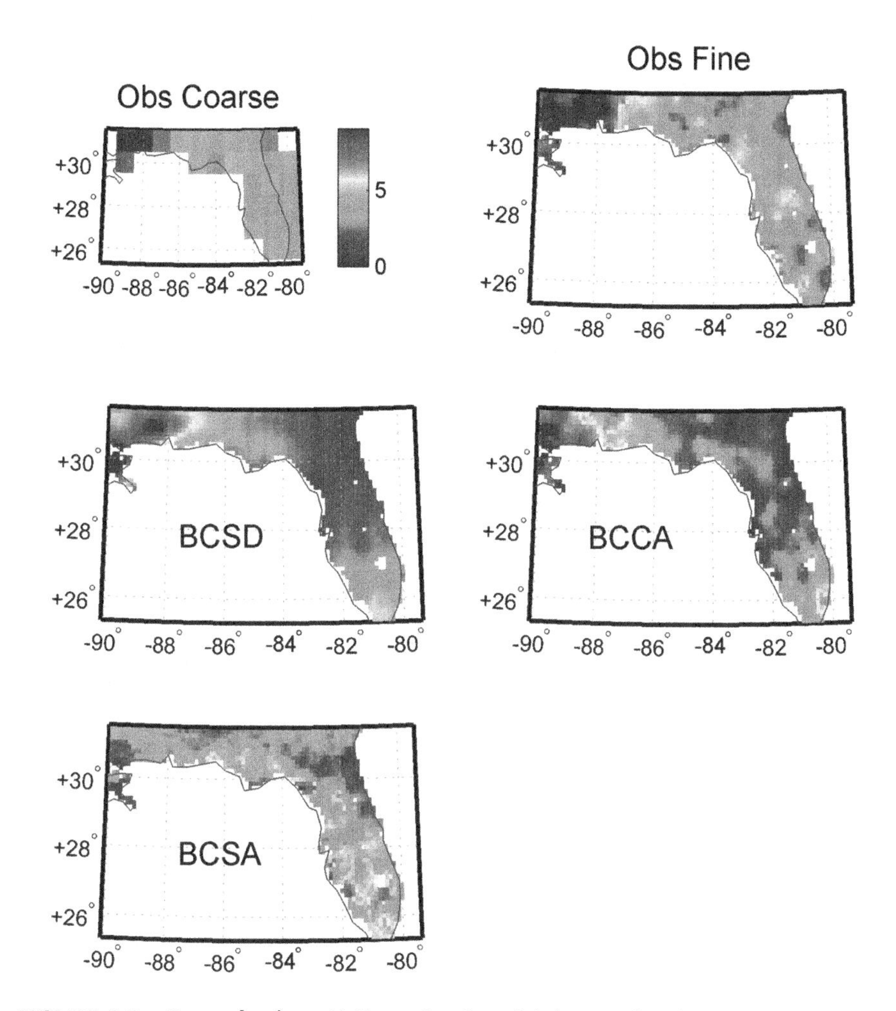

FIGURE 9.2 (See color insert.) Example of spatial downscaling for monthly precipitation in a mean daily rate for coarse- and fine-scale data for Florida on May, 1990 with three BCSD, BCCA, and BCSA methods. The coarse and fine spatial observations are shown in the top left and right panels, respectively.

References

Bardossy, A. and Plate, E., 1992. Space-time model for daily rainfall using atmospheric circulation patterns. *Water Resources Research*, 28(5): 1247–1259. doi:10.1029/91WR02589.

Brekke, L., Thrasher, B.L., Maurer, E.P. and Pruitt, T., 2013. *Downscaled CMIP3 and CMIP5 Climate and Hydrology Projections*. Bureau of Reclamation, Washington.

Caffrey, P. and Farmer, A., 2014. *A Review of Downscaling Methods for Climate Change Projections: African and Latin American Resilience to Climate Change (ARCC)*. United States Agency for International Development by Tetra Tech ARD.

Cannon, A.J., Sobie, S.R. and Murdock, T.Q., 2015. Bias correction of GCM precipitation by quantile mapping: How well do methods preserve changes in quantiles and extremes? *Journal of Climate*, 28(17): 6938–6959. doi:10.1175/jcli-d-14–00754.1.

Chen, J., Brissette, F.P. and Leconte, R., 2011. Uncertainty of downscaling method in quantifying the impact of climate change on hydrology. *Journal of Hydrology*, 401(3–4): 190–202. doi:10.1016/j.jhydrol.2011.02.020.

Cunnane, C., 1978. Unbiased plotting positions - A review. *Journal of Hydrology*, 37(3–4): 205–222. doi:10.1016/0022-1694(78)90017–3.

Devore, J.L., 1995. *Probability and Statistics for Engineering and the Sciences*. Duxbury Press, Oxford, 743 pp.

Dickinson, R.E., Errico, R.M., Giorgi, F. and Bates, G.T., 1989. A regional climate model for the western United States. *Climatic Change*, 15(3): 383–422. doi:10.1007/BF00240465.

Draper, N.R. and Smith, H., 1966. *Applied Regression Analysis*. Wiley, New York.

Efroymson, M.A., 1966. Stepwise regression—A backward and forward look Presented at the Eastern Regional Meetings of the Inst. of Math. Statist.

Feser, F., Rrockel, B., Storch, H., Winterfeldt, J. and Zahn, M., 2011. Regional climate models add value to global model data a review and selected examples. *Bulletin of the American Meteorological Society*, 92(9): 1181–1192. doi:10.1175/2011BAMS3061.1.

Flato, G. et al., 2013. Evaluation of climate models. In: *Climate Change 2013: The Physical Science Basis*, IPCC, Cambridge, UK, pp. 746–811.

Geem, Z.W., 2009. *Music-Inspired Harmony Search Algorithm: Theory and Applications*. Springer-Verlag Berlin Heidelberg, Chennai, 206 pp. doi:10.1007/978–3–642–00185–7.

Geem, Z.W., 2006. Optimal cost design of water distribution networks using harmony search. *Engineering Optimization*, 38(3): 259–280. doi:10.1080/03052150500467430.

Geem, Z.W., Kim, J.H. and Loganathan, G.V., 2002. Harmony search optimization: Application to pipe network design. *International Journal of Modelling and Simulation*, 22(2): 125–133.

Geem, Z.W., Kim, J.H. and Loganathan, G.V., 2001. A new heuristic optimization algorithm: Harmony search. *Simulation*, 76(2): 60–68.

Grillakis, M.G., Koutroulis, A.G. and Tsanis, I.K., 2013. Multisegment statistical bias correction of daily GCM precipitation output. *Journal of Geophysical Research: Atmospheres*, 118(8): 3150–3162. doi:10.1002/jgrd.50323.

Gumbel, E.J., 1954. *Statistical Theory of Extreme Values and Som Practical Applications*. Applied Mathematics Series, National Bureau of Standards, Washington D.C.

Hidalgo, H.G., Dettinger, M.D. and Cayan, D.R., 2008. Downscaling with constructed analogues: Daily precipitation and temperature fields over the United States, California Energy Commission, California Climate Change Center Report Series Number 2007-027.

Hwang, S. and Graham, W.D., 2013. Development and comparative evaluation of a stochastic analog method to downscale daily GCM precipitation. *Hydrology and Earth System Sciences*, 17(11): 4481–4502. doi:10.5194/hess-17–4481–2013.

IPCC, 2001. Climate Change 2001: The scientific basis: Contribution of Working Group I to the Third Assessment Report of the Intergovernmental Panel on Climate Change, Cambridge University Press, Cambridge.

Jenkinson, A.F. and Collison, F.P., 1977. An initial climatology of gales over the North Sea. Synoptic Climatology Branch Memorandum, Report No.62, Meteorological Office, Bracknell .

Johnson, R.A. and Wichern, D.W., 2001. *Applied Multivariate Statistical Analysis.* Prentice Hall, Upper Saddle River, 767 pp.

Jones, P. and Harpham, C., 2009. *UK Climate Projections Science Report: Projections of Future Daily Climate for the UK from the Weather Generator.* University of Newcastle, Newcastle.

Jones, P.D., Hulme, M. and Briffa, K.R., 1993. A comparison of Lamb circulation types with an objective classification scheme. *International Journal of Climatology*, 13(6): 655–663. doi:10.1002/joc.3370130606.

Kottegoda, N.T. and Rosso, R., 2008. *Applied Statistics for Civil and Environmental Engineers.* Wiley-Blackwell, West Sussex, UK, 736 pp.

Kottegoda, N.T. and Rosso, R., 1997. *Statistics, Probability, and Reliability for Civil and Environmental Engineers.* McGraw-Hill, New York, 735 pp.

Lall, U. and Sharma, A., 1996. A nearest neighbor bootstrap for resampling hydrologic time series. *Water Resources Research*, 32(3): 679–693.

Lamb, H.H., 1972. British Isles weather types and a register of the daily sequence of circulation patterns, 1861–1971. Meteorological Office, *Geophysical Memoirs*, No. 116: pp. 85.

Lee, A., Geem, Z.W. and Suh, K.D., 2016. Determination of optimal initial weights of an artificial neural network by using the harmony search algorithm: Application to break-water armor stones. *Applied Sciences (Switzerland)*, 6(6). doi:10.3390/app6060164.

Lee, T. and Jeong, C., 2014. Nonparametric statistical temporal downscaling of daily precipitation to hourly precipitation and implications for climate change scenarios. *Journal of Hydrology*, 510: 182–196. doi:10.1016/j.jhydrol.2013.12.027.

Lee, T. and Ouarda, T.B.M.J., 2011. Identification of model order and number of neighbors for k-nearest neighbor resampling. *Journal of Hydrology*, 404(3–4): 136–145. doi:10.1016/j.jhydrol.2011.04.024.

Lee, T., Ouarda, T.B.M.J. and Jeong, C., 2012. Nonparametric multivariate weather generator and an extreme value theory for bandwidth selection. *Journal of Hydrology*, 452–453: 161–171. doi:10.1016/j.jhydrol.2012.05.047.

Lee, T., Salas, J.D. and Prairie, J., 2010. An enhanced nonparametric streamflow disaggregation model with genetic algorithm. *Water Resources Research*, 46: W08545. doi:10.1029/2009WR007761.

Leloup, J., Lengaigne, M. and Boulanger, J.P., 2008. Twentieth century ENSO characteristics in the IPCC database. *Climate Dynamics*, 30(2–3): 277–291. doi:10.1007/s00382-007-0284–3.

Li, H., Sheffield, J. and Wood, E.F., 2010. Bias correction of monthly precipitation and temperature fields from Intergovernmental Panel on Climate Change AR4 models using equidistant quantile matching. *Journal of Geophysical Research: Atmospheres*, 115(10). doi:10.1029/2009JD012882.

Mackay, D.J.C., 1992. Bayesian interpolation. *Neural Computation*, 4(3): 415–447.

MacQueen, J., 1967. Some methods for classification and analysis of multivariate observations. *Proceedings of the Fifth Berkeley Symposium on Mathematical Statistics and Probability*, 1: 281–297.

Maraun, D., 2016. Bias correcting climate change simulations—A critical review. *Current Climate Change Reports*, 2(4): 211–220. doi:10.1007/s40641-016-0050–x.

Maraun, D., 2013. Bias correction, quantile mapping, and downscaling: Revisiting the inflation issue. *Journal of Climate*, 26(6): 2137–2143. doi:10.1175/JCLI-D–12–00821.1.

Maraun, D. et al., 2010. Precipitation downscaling under climate change: Recent developments to bridge the gap between dynamical models and the end user. *Reviews of Geophysics*, 48(3), RG3003 pp.1–34.

Matalas, N.C., 1967. Mathematical assessment of synthetic hydrology. *Water Resources Research*, 3(4): 937–945. doi:10.1029/WR003i004p00937.

Maurer, E.P. and Hidalgo, H.G., 2008. Utility of daily vs. monthly large-scale climate data: An intercomparison of two statistical downscaling methods. *Hydrology and Earth System Sciences*, 12(2): 551–563.

Maurer, E.P., Hidalgo, H.G., Das, T., Dettinger, M.D. and Cayan, D.R., 2010. The utility of daily large-scale climate data in the assessment of climate change impacts on daily streamflow in California. *Hydrology and Earth System Sciences*, 14(6): 1125–1138. doi:10.5194/hess-14–1125–2010.

McGuffie, K. and Henderson-Sellers, A., 2005. *A Climate Modelling Primer: Third Edition*. John Wiley and Sons, 1–280 pp. doi:10.1002/0470857617.

Nasseri, M., Asghari, K. and Abedini, M.J., 2008. Optimized scenario for rainfall forecasting using genetic algorithm coupled with artificial neural network. *Expert Systems with Applications*, 35(3): 1415–1421. doi:10.1016/j.eswa.2007.08.033.

Park, S.K. and Miller, K.W., 1988. Random number generators: Good ones are hard to find. *Communications of the ACM*, 31(10): 1192–1201. doi:10.1145/63039.63042.

Piani, C., Haerter, J.O. and Coppola, E., 2010. Statistical bias correction for daily precipitation in regional climate models over Europe. *Theoretical and Applied Climatology*, 99(1–2): 187–192. doi:10.1007/s00704-009-0134–9.

Press, W., Teukolsky, S., William, T. and Flannery, B., 2002. *Numerical Recipes in C++. The Art of Scientific Computing*. Cambridge University Press, New York.

Rajagopalan, B. and Lall, U., 1999. A k-nearest-neighbor simulator for daily precipitation and other weather variables. *Water Resources Research*, 35(10): 3089–3101.

Richardson, C.W. and Wright, D.A., 1984. *WGEN: A Model for Generating Daily Weather Variables*. Agricultural Research Service, US Department of Agriculture, Springfield, MA.

Rust, H.W., Vrac, M., Lengaigne, M. and Sultan, B., 2010. Quantifying differences in circulation patterns based on probabilistic models: IPCC AR4 multimodel comparison for the North Atlantic. *Journal of Climate*, 23(24): 6573–6589. doi:10.1175/2010JCLI3432.1.

Salathé Jr, E.P., Mote, P.W. and Wiley, M.W., 2007. Review of scenario selection and downscaling methods for the assessment of climate change impacts on hydrology in the United States pacific northwest. *International Journal of Climatology*, 27(12): 1611–1621.

Sedki, A. and Ouazar, D., 2010. Hybrid particle swarm and neural network approach for streamflow forecasting. *Mathematical Modelling of Natural Phenomena*, 5(7): 132–138. doi:10.1051/mmnp/20105722.

Sharif, M. and Burn, D.H., 2006. Simulating climate change scenarios using an improved K-nearest neighbor model. *Journal of Hydrology*, 325(1–4): 179–196.

Srinivasulu, S. and Jain, A., 2006. A comparative analysis of training methods for artificial neural network rainfall–runoff models. *Applied Soft Computing*, 6(3): 295–306. doi:10.1016/j.asoc.2005.02.002.

Storch, H.v., Zorita, E. and Cubasch, U., 1993. Downscaling of global climate change estimates to regional scales: An application to iberian rainfall in wintertime. *Journal of Climate*, 6(6): 1161–1171. doi:10.1175/1520-0442(1993)006<1161:dogcce>2.0.co;2.

Sturges, H.A., 1926. The choice of a class interval. *Journal of the American Statistical Association*, 21(153): 65–66. doi:10.1080/01621459.1926.10502161.

Thrasher, B., Maurer, E.P., McKellar, C. and Duffy, P.B., 2012. Technical note: Bias correcting climate model simulated daily temperature extremes with quantile mapping. *Hydrology and Earth System Sciences*, 16(9): 3309–3314. doi:10.5194/hess-16–3309–2012.

Tibshirani, R., 1996. Regression shrinkage and selection via the lasso. *Journal of the Royal Statistical Society. Series B (Methodological)*, 58(1): 267–288.

Van Brummelen, G., 2012. *Heavenly Mathematics: The Forgotten Art of Spherical Trigonometry*. Princeton University Press, Princeton, 208 pp.

Washington, W.M. and Parkinson, C.L., 2005. *An Introduction to Three-Dimensional Climate Modeling*. University Science Books, Sausalito, CA, 354 pp.

Wasserman, P.D., 1993. *Advanced Methods in Neural Computing*. John Wiley & Sons, New York.

Watson, R.T., Zinyowera, M.C. and Moss, R.H., 1996. *Climate Change 1995—Impacts, Adaptations and Mitigation of Climate Change: Scientific-Technical Analyses*. Cambridge University Press, Cambridge, UK, 879 pp.

Wilby, R.L., 1994. Stochastic weather type simulation for regional climate change impact assessment. *Water Resources Research*, 30(12): 3395–3402.

Wilby, R.L., Dawson, C.W. and Barrow, E.M., 2002. SDSM—A decision support tool for the assessment of regional climate change impacts. *Environmental Modelling & Software*, 17(2): 145–157. doi:10.1016/S1364-8152(01)00060-3.

Wilby, R.L. et al., 1998. Statistical downscaling of general circulation model output: A comparison of methods. *Water Resources Research*, 34(11): 2995–3008.

Yates, D., Gangopadhyay, S., Rajagopalan, B. and Strzepek, K., 2003. A technique for generating regional climate scenarios using a nearest-neighbor algorithm. *Water Resources Research*, 39(7): 1199. doi:10.1029/2002WR001769.

Yoon, S., Jeong, C. and Lee, T., 2013. Application of harmony search to design storm estimation from probability distribution models. *Journal of Applied Mathematics*, 2013. doi:10.1155/2013/932943.

Young, K.C., 1994. A multivariate chain model for simulating climatic parameters from daily data. *Journal of Applied Meteorology*, 33(6): 661–671.

Index

L

M

N

O

P

Q

R

S

T

U

W